Performance Improvement in Construction Management

Spon Research

Spon Research publishes a stream of advanced books for built environment researchers and professionals from one of the world's leading publishers. The ISSN for the Spon Research programme is ISSN 1940–7653 and the ISSN for the Spon Research E-book programme is ISSN 1940–8005

Published

Free-Standing Tension Structures
From tensegrity systems to
cable-strut systems
978-0-415-33595-9
W.B. Bing

Performance-Based Optimization of Structures
Theory and applications
978-0-415-33594-2
Q.Q. Liang

Microstructure of Smectite Clays and Engineering Performance
978-0-415-36863-6
R. Pusch and R. Yong

Procurement in the Construction Industry
The impact and cost of alternative
market and supply processes
978-0-415-39560-1
W. Hughes et al.

Communication in Construction Teams
978-0-415-36619-9
S. Emmitt and C. Gorse

Concurrent Engineering in Construction Projects
978-0-415-39488-8
C. Anumba, J. Kamara and A.-F. Cutting-Decelle

People and Culture in Construction
978-0-415-34870-6
A. Dainty, S. Green and B. Bagilhole

Very Large Floating Structures
978-0-415-41953-6
C.M. Wang, E. Watanabe and T. Utsunomiya

Tropical Urban Heat Islands
Climate, buildings and greenery
978-0-415-41104-2
N.H. Wong and C. Yu

Innovation in Small Construction Firms
978-0-415-39390-4
P. Barrett, M. Sexton and A. Lee

Construction Supply Chain Economics
978-0-415-40971-1
K. London

Employee Resourcing in the Construction Industry
978-0-415-37163-6
A. Raiden, A. Dainty and R. Neale

Managing Knowledge in the
Construction Industry
978-0-415-46344-7
A. Styhre

Collaborative Information
Management in Construction
978-0-415-48422-0
*G. Shen, A. Baldwin and
P. Brandon*

Forthcoming

Organisational Culture in the
Construction Industry
978-0-415-42594-0
V. Coffey

Containment of High Level
Radioactive and Hazardous Solid
Wastes with Clay Barriers
978-0-415-45820-7
*R.N. Yong, R. Pusch and
M. Nakano*

Relational Contracting for
Construction Excellence
Principles, practices and partnering
978-0-415-46669-1
A. Chan, D. Chan and J. Yeung

Performance Improvement in Construction Management

Edited by Brian Atkin and
Jan Borgbrant

Routledge
Taylor & Francis Group

LONDON AND NEW YORK

First published 2010
by Spon Press
2 Park Square, Milton Park, Abingdon, Oxon OX14 4RN

Simultaneously published in the USA and Canada by Spon Press
711 Third Avenue, New York, NY 10017

Routledge is an imprint of the Taylor & Francis Group, an informa business

First issued in paperback 2016

Typeset in Sabon by
Wearset Ltd, Boldon, Tyne and Wear
Printed and bound in Great Britain by
CPI Antony Rowe, Chippenham, Wiltshire

British Library Cataloguing in Publication Data
A catalogue record for this book is available from the British Library

Library of Congress Cataloging-in-Publication Data
Performance improvement in construction management / edited by
Brian Atkin and Jan Borgbrant.
p. cm. – (Spon research, ISSN 1940-7653)
Includes bibliographical references and index.
1. Building–Superintendence. 2. Construction industry–Management. 3.
Industrial productivity. I. Atkin, Brian. II. Borgbrant, Jan.
TH438.P384 2009

690.068–dc22   2009015344

ISBN13: 978-0-415-54598-3 (hbk)
ISBN13: 978-1-138-97818-8 (pbk)

Contents

Preface

Our motivation for preparing this collection of research results is the need to bring together a body of further understanding and insights into the ways we think about the economics and organization of construction – broadly speaking, the field of construction management. The recurrent theme of this book, as reflected in its 24 contributions, is questioning traditional (i.e. linear) thinking and practices and how these impact on the construction process in general. Our interest is, however, much wider and more forward looking, but we believe it is important to understand some of the fundamental weaknesses in the way that the construction industry is organized and how it accounts for itself to owners and clients. We have also tried to ensure that focus remains on the end product's performance and other dimensions of quality, as well as cost.

Performance Improvement in Construction Management grew out of the latest in a series of international conferences – the 4th Nordic Conference on Construction Economics and Organization – in Luleå, Sweden and a national research program, *Competitive Building*. The 50 contributors are a cross-section of the research community, representing more than ten countries and a substantial body of research activity. The output of their research stems from, in many cases, longstanding collaboration between researchers and industry partners. Some of the contributors have a dual role as researchers from companies working collaboratively in internationally-leading universities. Their role can be described as that of *change agent*, where the problem identified in-company will be resolved by the implementation of research results in-company.

This book draws together research results in three areas: innovative processes; organization and human behavior; and managerial methods, techniques and tools. Each area contains a cluster of chapters that describe and map out new approaches for industry practitioners, based on solid research supported in many cases by leading companies and other organizations acting as a proxy for industry's interests. Working at the interface between known problems in the field and new understanding of processes, behavior, methods and tools, *Performance Improvement in Construction Management* reveals progress made in a number of collaborative projects between industry and the research community. We believe that a particular strength of the book is its transparency in revealing the source of ideas, data and results. If the construction industry is to improve its performance by becoming more competitive and efficient, the actors within it need to be able to recognize credible examples and the steps required to achieve sustainable improvement.

Brian Atkin and Jan Borgbrant
Reading and Stockholm

Contributors

Brian Atkin, Visiting Professor, Lund University and University of Reading.

Reza Beheshti, Delft University of Technology.

Niels Haldor Bertelsen, Danish Building Research Institute.

Johan Björnström, Chalmers University of Technology.

Jan Borgbrant, Professor Emeritus, Luleå University of Technology.

Birgit Brunklaus, Chalmers University of Technology.

Nicola Costantino, Politecnico di Bari.

Hans de Jonge, Delft University of Technology.

Hennes de Ridder, Delft University of Technology.

Giuseppe Dibari, Politecnico di Bari.

André Dorée, University of Twente.

Marco Dreschler, Delft University of Technology.

Michael Edén, Chalmers University of Technology.

Ellen Gehner, Delft University of Technology.

Jac L.A. Geurts, University of Tilburg.

Pernilla Gluch, Chalmers University of Technology.

Cecilia Gustafsson, Chalmers University of Technology.

Karin Johansson, Chalmers University of Technology.

Rasmus Johansson, Lund University.

Anna Kadefors, Chalmers University of Technology.

Kalle Kähkönen, Technical Research Centre of Finland.

Kari Hovin Kjølle, Norwegian University of Science and Technology.

Albertus Laan, University of Twente.

Anne Landin, Lund University.

Erika Levander, Luleå University of Technology.

Alexander Löfgren, The Royal Institute of Technology.

Louis Lousberg, Delft University of Technology.

Örjan Lundberg, Chalmers University of Technology.

Seirgei Miller, The Cape Peninsula University of Technology.

Björn Niklasson, Lund University.

Stefan Olander, Lund University.

Ekaterina Osipova, Luleå University of Technology.

Gert-Joost Peek, ING Real Estate and Delft University of Technology.

Mats Persson, Lund University.

Roberto Pietroforte, Worcester Polytechnic Institute.

Christine Räisänen, Chalmers University of Technology.

Steve Rowlinson, University of Hong Kong.

Barbara Rubino, Chalmers University of Technology.

Silvio Sancilio, Politecnico di Bari.

Jutta Schade, Luleå University of Technology.

Kajsa Simu, Luleå University of Technology.

Lars Stehn, Luleå University of Technology.

Ann-Charlotte Stenberg, Chalmers University of Technology.

Henny ter Huerne, University of Twente

Liane Thuvander, Chalmers University of Technology.

Martin Tuuli, University of Hong Kong.

Anders Vennström, Luleå University of Technology.

Ruben Vrijhoef, Delft University of Technology.

Hans Wamelink, Delft University of Technology.

Abukar Warsame, The Royal Institute of Technology.

Örjan Wikforss, The Royal Institute of Technology.

Tas Koh Yong, University of Hong Kong.

Abbreviations

BCIS	Building Cost Information Service
CIO	Chief Information Officer
EMAT	economically most advantageous tender
ER	Engineer's Representative
GIS	geographic information system
GMP	guaranteed maximum price
GPS	global positioning system
ICT	information and communications technology
ISO	International Organization for Standardization
LCC	life cycle cost
LCCA	life cycle cost analysis
NPV	net present value
SME	small- and medium-sized enterprise
TCE	transaction cost economics
TTA	temporary traffic arrangement
TVE	timber volume element
WLA	whole life appraisal
WLC	whole life costing

1 Steps toward improving performance in construction management

Jan Borgbrant and Brian Atkin

Introduction

The field of economics and organization in construction – broadly speaking, construction management – provides the framework for this book, which is a collection of research results clustered in three areas:

1 innovative processes;
2 organization and human behavior; and
3 managerial methods, techniques and tools.

The results presented here are linked to the field of construction management and are reflective of different practices in the countries in which the contributors are working. While all have academic affiliations, most have strong connections with industry. In many cases, the results represent the outcomes of research collaboration between universities and companies; in some cases, the relationship has been longstanding such that the results represent the more recent contributions.

A theme running through this book is questioning traditional (i.e. linear) thinking and practices and how these impact the construction process in general. A linear way of thinking and acting is one that is based on the typical steps taken in the construction process. These steps follow a logical, technical structure – from concept to operations – and apply to almost all forms of construction. We can consider this approach from three perspectives in relation to the field of construction:

1 actors
2 developments
3 research and researchers.

Actors in the field

The different actors in the construction field are shown in Figure 1.1. In the center circle, we find end-users of the products that are created, for example housing, industrial and commercial buildings as well as various kinds of infrastructure. Between the circles are groups of actors. These groups participate in different phases and stages of the construction process. Sometimes, a construction company enters into discussion with a client's representative; on other occasions, the first point of contact for the client's representative will be a designer. The steps that follow involve suppliers and construction companies possessing different specializations, combining in a value chain to deliver the assets of the owner or client.

If we examine this position closely, we can say that:

- the motives and drivers of the many actors can vary significantly;
- actors have their own visions, aims and goals;
- companies or other organizations behind them have their own interests;
- organizations differ much in size from just a few to thousands of workers; and
- different customs, practices, norms and ethics are involved.

When focusing on the construction field some crucial questions arise, for example:

- how does communication between the actors actually occur?
- how are end-users' needs and requirements for the product expressed, communicated and understood among the actors in the construction process?
- are relevant actors active at the appropriate phase and stage in the construction process?

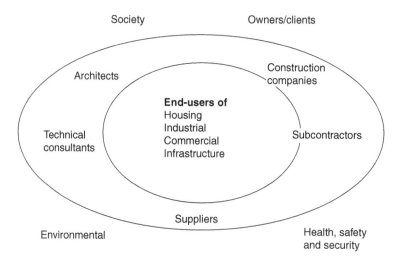

Figure 1.1 The construction field.

The results presented by the different contributors in this book can help to answer some of these questions and contribute to a better understanding of the rest.

Development of the field

A key question is the extent to which the construction field can be developed. This question is crucial from the perspective of the many actors in the construction process, each with their own knowledge, skills, experiences, attitudes and behavior. Each is generally seeking to maximize the gains from the process. The question can be discussed on three organizational levels: strategic, tactical and operational.

Strategic level

In looking at the construction field on the strategic level, we consider that analysis should be undertaken from two perspectives. Figure 1.2 shows these perspectives within a framework for change.

Experience of 'what the construction process is like today' and perceptions of 'how it should be' differ much among the actors. If any changes are to occur, it is necessary to allow the actors to be a part of the change process so they can reach a common understanding of 'what needs to be changed' and 'how to make those changes sustainable'. It is important to observe factors that support change and factors that represent barriers to change.

Tactical level

When the need for change has been identified, understood and, hopefully, accepted by all involved, progress can be made on moving toward the desired changes drawing on various means or methods. The three primary means are shown in Figure 1.3.

These three means or methods are closely connected. The concept of education has an organizational structure and this can be in the form of

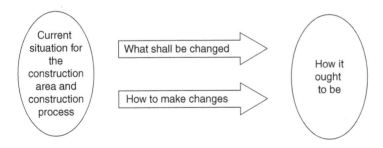

Figure 1.2 Framework for change.

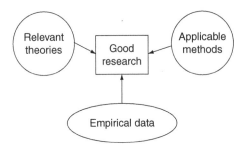

Figure 1.4 Building blocks of *good* research.

research process. One of the most important issues for the researcher is applying the results within companies in the field.

Contributions to the field of research

Our discussions and illustrations may appear simple and straightforward – they are intended to be – because they are the models that underpin the research discussed in this book and the results that have emerged from it. In this section, we summarize the contributions in each chapter.

Innovative processes

Anders Vennström (Chapter 2) investigates how construction clients handle the transformation of requirements in the early stage of the construction process in order to contribute to change in that process. Since the early stage of the construction process defines the mission of the project, the client should therefore scrutinize the present situation in order to correctly formulate the mission for the project. In this way, the construction client can be instrumental as an initiator of change. The research reveals that clients' decision-making processes were, however, more concerned with management (planning, budgeting, controlling and problem solving) of the process than leadership or driving change. They frequently referred to basic data for decisions that were linked to operational matters. The strategic issues for the construction process were not taken into consideration to the same extent. In order to act as a change agent, clients need to become an integrator of requirements from the various stakeholders in the project.

Stefan Olander *et al.* (Chapter 3) had the aim of identifying costs arising from the impact of external factors in the planning and development process leading up to final approval by the local authority. The research focuses on direct costs resulting in real payments; even so, indirect costs in the form of interest rates and delayed revenues are also relevant and are discussed. Two approaches to the engagement of external stakeholders are examined and from them an appropriate strategy for development projects is proposed. The management of the planning process can be improved by streamlining the early planning stages through identifying the key issues. Improvement here should give both the local authority and the developer common ground for future decision-making in the development process. These actions should decrease the uncertainty related to external impacts on the planning process. As one of the primary stakeholders, the authority has the opportunity to become a positive force in real estate development rather than a constraint.

Ruben Vrijhoef and Hennes de Ridder (Chapter 4) adopt an approach where a construction supply chain is regarded as a system and then apply systems engineering to increase the coherence of that system. The underlying principle is that a production system such as a supply chain delivering a single product should not be fragmented, neither should it consist of distributed functions. Instead, supply chain integration must lead to improvement by developing a more stable, repetitive production environment, similar to that found in other industries. The premise is that the construction supply chain would function better when approached and reconstituted as a single entity in the form of an extended enterprise. In a way, the broader issue is whether construction could or should develop itself toward the standards and practices of a conventional, more integrated, supply-driven industry. The research offers insights into a number of the *building blocks* found in theory and practice, which can be used in the model building process.

Örjan Wikforss and Alexander Löfgren (Chapter 5) argue that R&D initiatives in the industry have been dominated by the purely technological development of information and communications technology (ICT). These efforts have not resulted in a comprehensive understanding of how new technology works in project communication if considering human, organizational and process-related factors, as well as those of a purely technological nature. Some of the fundamental collaborative communication issues in the planning, design and production phases of construction projects are described. An initial, conceptual framework for developing communication practices, combined with supportive ICT as a facilitator for improved organization and management of future construction projects, is outlined. The research concludes that full recognition of the need and determination to improve collaborative communication and information exchange throughout all project phases could have considerable impact on the industrialization of construction projects.

Mats Persson and Anne Landin (Chapter 6) press the case for a system of reporting experience gained on projects that is designed in such a way so all engaged in the chain of tasks to be performed can access knowledge of the experience of others. If no such transfer of knowledge and experience takes place, there is the two-fold risk of the firm failing to take advantage of what has been learnt and of its making similar mistakes again. The complexity of the construction process means, however, that special measures are called for if the collection of relevant information, including that concerned with experience gained, is to fulfill its purpose. The continual public debate regarding what takes place within the industry is considered to reflect flaws in quality assurance systems and the lack of well-functioning systems for collecting and distributing knowledge, which empirical study confirms. There is good reason, therefore, for the industry to identify ways in which the functioning of these two systems can be improved.

Niels Haldor Bertelsen (Chapter 7) highlights the enormous changes occurring in the industry and uses production as an example of the shift toward industrial methods. New forms of partnership and models of organization are also being introduced in and between enterprises. A multitude of new materials and products are replacing old familiar ones, and the industry is under pressure to use digital models and 3D visualization in communicating with end-users and collaboration partners. How do specialist contractors and small firms respond to these challenges? What is the role of the skilled worker in this development? How can firms and workers contribute to a solution to these problems and improve their position and opportunities for development? The research is a first step, aimed at finding a structure and model for efforts to increase competence in innovation. The results are encouraging, such that a target group of small construction firms and skilled workers is able to carry out accelerated development in different modes toward the goal of industrialized production.

Louis Lousberg and Hans Wamelink (Chapter 8) found that in the theoretical foundation of project management the product approach still outweighs the process approach, wherein the latter deals with power and politics. The choice of the *linguistic turn*, as a starting point of the paradigm change that has been made in organizational studies, makes that approach more of a process. The foundations, the theorizing and examples of these studies are explored in this chapter. The exploration is based on a literature study and preliminary results of research into conflicts in complex public–private projects. First, the ontological and epistemological starting point of this approach is elaborated by a description of an attempt to escape the subject/object dualism. Next, the following theses are developed:

1 there is not one theory, but there are multiple approaches;
2 there is not one form of managing projects, but there are several; and
3 due to the gap of knowledge of social project management forms research should be focused on practice from an interpretative perspective.

Seirgei Miller *et al.* (Chapter 9) describe an initiative aimed at improving quality in the construction process. The road construction sector offers valuable insights into how changes are reshaping what have been long-standing ways of working. This chapter examines one important aspect of road construction – the asphalt paving process. In an effort to outperform competitors, asphalt paving companies are seeking better control over the paving process, over the planning and scheduling of resources and work, and over performance. Improved control can also reduce the risks of failure of the paving during the period of guarantee. To be able to achieve these goals, the relevant operational parameters need to be known and the relationships between these parameters have to be thoroughly understood. For asphalt paving companies to be able to improve both product and process performance, they have to develop a more sophisticated understanding of the asphalt paving process and the interdependencies within it.

Organization and human behavior

Johan Björnström *et al.* (Chapter 10) question the meaning of communication in a large company. Based on a longitudinal case study of the company's strategy development during a change process, the chapter focuses on the ways in which the new strategies were communicated and interpreted at four organizational levels. Theories from literature on strategy were used to provide a socio-cultural lens with which to view the findings. Different perspectives on the function of organizational strategies, as well as on the meaning of the term communication, were found and these gave rise to different managerial approaches to strategy development. This in turn led to different values being ascribed to the strategy process, which consequently gave rise to different interpretations of the strategies themselves. This chapter shows that communication is a complex and dynamic, socially constructed process which is inherently intertwined with other organizational activities. It needs therefore to be continuously discussed and evaluated in the same way as other processes and activities.

Abukar Warsame (Chapter 11) contends that traditional organizational structure studies emphasize department groupings and the management style of different organizations which are often based on common tasks, products, geography and processes. This chapter adopts a different tack, by analyzing various models of the organization of the firm within house-building from the perspective of transaction cost theory. As a method of general evaluation of different organizational structures, three criteria were chosen as they are in line with the transaction cost approach: flexibility and risk allocation, competition and competence. The use of transaction cost theory as a tool for exploring different organizational structures from an efficiency perspective makes it easier to predict how firms engaged in projects respond to economic and business challenges that are vital for their survival. It may explain why particular organizational structures have dominated at some point. It also helps to reveal forces (competitive pressure, higher level of required competence, greater flexibility etc.) that make it necessary to consider another organizational form.

Anna Kadefors and Albertus Laan (Chapter 12) examine levels of trust in inter-organizational cooperation. In industry, concerns are often raised that higher levels of trust would improve performance. Trust is a complex and multidimensional construct, involving conscious calculation as well as emotions and intuition. Processes on the individual, organizational and societal levels interact in shaping trust. In construction, temporary and unique project organizations place high demands on information processing. Extensive industry-level standardization of roles and procurement routes has been developed, while the amount of face-to-face mutual adjustment has been kept down. Trust production in construction is characterized by a strong emphasis on institutional trust (thin trust), while relational trust (thick trust) is neglected. In this weak trust context contracts influence trust negatively, since changes tend to produce tensions. To improve collaboration, more resources need to be spent on project-level communication. Also, industry-level standardization should be better adapted to goals of flexibility and joint explorative learning.

Marco Dreschler *et al.* (Chapter 13) criticize the dominant strategy of certain public clients over their persistence with procurement based on lowest price. Companies are forced to customize beyond the economic optimum, causing the industry to underperform. As a result of the lowest price-based procurement regime, contractors and, indeed, the entire supply chain have to comply with a detailed list of demands, setting up a new production system each time, which means the supply chain cannot organize itself in the most economical way. Bidding regulations allow value-based procurement, but this criterion is difficult to implement. Several value-based procurement methods have been delayed by legal claims, due to errors in the award mechanism. Nonetheless, examples of successful implementation can be found. The study presented in this chapter identifies the distinguishing features that influence acceptability and soundness, by analyzing applications of the EMAT award mechanism. Findings based on five cases where EMAT has been applied are presented.

Karin Johansson (Chapter 14) points out that mergers are generally seen as part of a strategy designed to achieve corporate growth. In practice, employees can experience confusion and sometimes resentment, hence the risk that they show less commitment or even disloyalty to the new company. As a result, productivity can decrease and in the long run these actions will show up in turnover of employees. Studies reveal that most failures in merging two companies result from poorly managed cultural integration. A sound integration should consider cultural and organizational aspects: a vital tool in these processes is communication. By undertaking interviews and analyzing literature on organizational culture and mergers, the chapter examines the merging of two mid-sized construction companies. The results indicate that cultural differences between merging organizations need to be managed in order to support the development of a common organizational culture. In this process, the structure and content of communication in the organization are vital concerns.

Pernilla Gluch *et al.* (Chapter 15) examine the problem of the slow take-up of a stronger environmental focus in the industry. Information campaigns have raised general environmental awareness; yet, mainstream building practices do not seem to have undergone any marked changes. This raises the question of how environmental issues are actually dealt with in practice. Has development stagnated and, if so, why? What is causing green innovation inertia? Simply, what makes it slow? This chapter provides some answers to these questions by empirically examining environmental attitudes, management and performance in the industry. The chapter detects possible causes of deficiencies and reveals why development, despite much effort, sometimes does not go in the direction intended by senior managers. By focusing on relations between the definition of environmental challenge, measures adopted and the results of those measures, the chapter identifies trends and institutionalizing processes that hinder sustainable development.

Steve Rowlinson *et al.* (Chapter 16) examine the implications of the growing awareness of corporate social responsibility in company boardrooms and the realization that the public in general and interest groups in particular have become much more challenging of authorities, institutions and corporations. The conditions under which construction projects are undertaken are conducive to disputes and hostilities from stakeholders. Not surprisingly, stakeholders are demanding that they be consulted and involved in developments of a construction nature. The response from industry is to use corporate social responsibility as a business strategy and as a driver of stakeholder engagement and management. 'Respect for people' is, as a consequence, becoming a core theme in construction organizations. Against this background the issue of relationship management has become prominent in stakeholder management discourse. The chapter examines through the lens of two cases studies the means by which relationship management can shape stakeholder management.

Barbara Rubino and Michael Edén (Chapter 17) explore the process by which novel technology is implemented within project teams. The interest in their investigation is focused on how energy-efficiency goals in building affect the current practice of a project team. The chapter adopts a cognitive approach and looks at the project team as a social setting and as the fundamental organization of construction enterprises. It is assumed to be the core of individual, firm and organizational learning, but also acts as the link where the possibilities for the introduction of innovations in buildings are tested, proposed or discarded. It is a dynamic system where learning goes on continuously, individually and collectively. But how and what do professionals learn within projects? Where are these goals more appropriately introduced in the design process? Simultaneously, many efforts are under way and the process is split amongst practitioners and projects. Some are based on earlier attempts; others start from scratch. Some use advanced and expensive components, while others try to simplify the technical solutions and choice of systems.

Managerial methods, techniques and tools

Gert-Joost Peek and Jac L.A. Geurts (Chapter 18) discuss the particular problem of city redevelopments, which tend to be characterized by long and laborious processes. The cause rests primarily in the concept phase of these processes, when objectives tend to be piled on top of each other on the road to ambitious, yet unrealistic, plans. At the start of these redevelopment processes, adequate knowledge and experience in dealing with the design commission are lacking. This leads to an unsteady course of actions and lowers the chances of enriching the design result. The research provides a new approach consisting of two parts to overcome these problems. One is a participative modeling exercise from which the actors learn about their own position in relation to the others in terms of how they value the redevelopment and the extent to which their ambitions are shared. The second is a participative design exercise that challenges the actors to communicate their wishes and ideas and to negotiate their objectives and willingness to invest. The method has been tested in practice and the participants have evaluated it as useful.

Kari Hovin Kjølle and Cecilia Gustafsson (Chapter 19) examine aspects of the design process for new offices, in particular, implementing clients' needs and meeting expectations. The process is heterogeneous, with several actors involved, e.g. architects, interior designers, engineers and clients. The different viewpoints imply a need for cooperation, often through negotiation. The tension that arises has to be managed if the process is to lead to a successful outcome. A case study is used to present some instruments that can facilitate the negotiations and the translation process between the clients on the demand side and architects and other designers on the supply side. In order to analyze how the collective action is managed across these social worlds, a theoretical framework from the field of science, technology and society has been utilized. The result is a focus on the concept defined as boundary objects, which are flexible, structured objects, such as analytical concepts, techniques and artifacts that help to establish a better and common understanding.

Erika Levander *et al.* (Chapter 20) discuss the limited application of LCC in construction from the perspective of owners and clients and identify the advantages and disadvantages of the main theoretical economic evaluation methods for LCC calculation, as well as the sources of relevant data. The results of a study of owners are presented, where the objectives were to identify uncertainties surrounding the greater adoption of timber-framed housing – specifically, TVE prefabrication – an increasingly common form of industrialized production. LCC was examined to see if it could be used to address those uncertainties. One conclusion is that LCC calculations are able to address a large number of the perceived uncertainties in relation to timber-framed housing. The model must, however, be broadened to include all aspects in integrated life cycle design for the analysis to be applicable and acceptable. Furthermore, the characteristics of timber-framed housing are such that it is crucial to address the uncertainties about its technical merits for it to be trusted by clients.

Ellen Gehner and Hans de Jonge (Chapter 21) argue that managing a real estate development project is really about risk taking. Yet, techniques of risk analysis are little used in the real estate development sector for reasons that include the paucity of objective data, time constraints and the lack of confidence in outcomes. Other methods are apparently used with success to manage risks in a project. For instance, investment decisions are generally presented as strategic questions. They are about allocating financial resources to a development project and, as a consequence, accepting the risks that result from this commitment. As risk is the outcome of a decision, decision making and risk management are intertwined. In order to get a better understanding of the way real estate developers treat risk, the decision process of one such developer is analyzed against six strategic decision process characteristics. From the analysis, it was concluded that a series of risk management strategies is embodied in the decision-making process. Moreover, factors that can help to explain organizational risk behavior do exist.

Ekaterina Osipova and Kajsa Simu (Chapter 22) investigate the risk management process on construction projects from the perspective of the client, contractor and consultant in general and in the context of small projects in particular. Preliminary research concluded that little attention is paid to identifying the roles of individual actors in risk management across a project's phases. No research seems to have been directed at small projects. The ways and extent to which the main actors are involved in risk management across the different phases of projects is examined and practices compared with the generally received view of risk management. Amongst the findings is the concern that consultants (typically architects and engineers in the design phase) have a low understanding of, and interest in, risk management. The particular case of small projects is given closer scrutiny in order to understand the extent to which systematic risk management takes place. The methods and tools used in this work are identified and are found to exclude formalized risk management systems.

Kalle Kähkönen (Chapter 23) argues for a quantitative approach to risk management which can provide an explicit link between the performance of projects and the financial wellbeing of the organizations responsible for their management. A major challenge is the necessary integration of the probabilistic and deterministic worlds. The need for integration can be regarded as a form of pressure intended to present risk management procedures with, and alongside, standard business functions. On the other hand, risk management as a separate function is a force that is pulling it away from other business functions. These two opposing forces, together with solutions that are highly situation-specific, need to be recognized as characteristics of risk management. This chapter presents and discusses a set of key elements that seem to form the basis for workable solutions of quantitative risk management in construction. Findings from three case studies are presented. In those cases, risk management procedures and related tools have been developed and implemented for use on live construction projects.

Silvio Sancilio *et al.* (Chapter 24) discuss barriers to entry in construction with particular emphasis on market concentration ratios. By analyzing the relationship between concentration ratio and market size, three market size categories can be defined: small, medium and large. Sweden, Italy and the USA were selected as the respective examples for these categories. Further analysis suggests that the concentration ratio is inversely proportional to market size. The analysis builds upon homogeneous data in order to make a consistent comparison among the three countries. The issue of subcontracting underscores the point that a construction market may be characterized by an apparent concentration (few main contractors) and, at the same time, fragmentation because of the presence of many subcontractors or specialist trade contractors. Within the US construction market there are sectors whose entry requires significant capital investments and/or know-how. These sectors are likely to have higher concentration ratios.

The 23 chapters that constitute the main body of research output here are themselves the products of a considerable body of knowledge as can be seen from the reference section at the back of this book.

2 Clients as initiators of change

Anders Vennström

Introduction

The construction industry has long been criticized over relationship related issues such as damaging conflicts and disputes, poor collaboration, and lack of customer focus and end-user involvement (Latham, 1994; Egan, 1998; Ericsson and Johansson, 1994). Researchers, practitioners and society at large have argued that changes in attitudes, behavior and procedures are necessary in order to increase the chances of project success and an improved end product (Love *et al.*, 2000; Dubois and Gadde, 2002). Two major investigations of the Swedish construction industry have pinpointed the role of the construction client as a 'driver for change' (Yngvesson *et al.*, 2000; Ericsson *et al.*, 2002). This is because construction clients occupy a key position regarding choice of procurement method and managerial processes, and clients set the basis for the governance of construction projects. For the purpose of this chapter, the construction client can be defined as 'the party that carries out or assigns others to carry out construction and related works'. In this connection, the client's role can be summarized as 'responsibility for interpreting and translating users' needs, expectations and desires into requirements and prerequisites for a project taking account of society's need for a sustainable built environment'.

The construction client is responsible for ensuring that the requirements of the owner(s), customers/end-users and society are met by a construction project at every point from concept to execution and handover. It is the construction client that purchases products and services from the construction industry, steers the construction process overall and thereby creates the conditions for the use and management of the building or facility over a long period.

Since construction is a complex, multi-discipline affair, where many actors come and go in the process, initiatives from one actor may not result in changes in other actors' behavior and procedures; indeed, there are arguments why this should not be allowed. Nonetheless, a change process, which is what a construction project represents, needs a clear definition of the present situation (problem formulation) that can then be shared among the participants in the process (i.e. the need for broad participation) (Borgbrant, 1990; Kotter, 1995; Tichy, 1983) in order to form visions and goals.

The construction process manifests as a temporary organization for practical reasons. If two identical buildings were produced in different places, at different times and by different organizations, the construction process would be a process unique to those circumstances (Kadefors, 1997). This kind of project arrangement becomes a venue for different experiences, interests, norms and ways of doing things: it becomes a scene for different organizational processes (Sahlin-Andersson, 1986).

The purpose of this chapter is to investigate how construction clients handle the transformation of requirements in the early stages of the construction process in order to contribute to change in that process. Since the early stages of the construction process define the mission of the project, the client should therefore scrutinize the present situation in order to correctly formulate the mission for the project (Ryd, 2003). In this way, the construction client can be instrumental as an 'initiator of change'.

Literature review

In this brief literature review, potentially relevant decision factors are examined. The literature review begins with a discussion of the main aspects and the issues involved.

Areas of need for the construction client

Decision-making in the early stages of the construction process is crucial for the client. It is there that intentions for the project are defined (Ryd, 2003). There are four main aspects – owner, customer/user, society and construction industry – for the client to handle and this must continue throughout the construction process in order to obtain the best result (Johansson and Svedinger, 1997).

The *owner* determines the *values* and *economic* demands to which the client has to relate. They can be formulated in *goals* and *visions* for the organization to which the client belongs (Johansson and Svedinger, 1997; Ljung, 1998). *Society* affects clients' decisions by virtue of *laws, regulations* and *opinions* (Cassel and Hjelmfeldt, 2001; Johansson and Svedinger, 1997). Restrictions and rights over property are crucial issues for clients to recognize and to deal with satisfactorily (Carn *et al.*, 1998). In particular, environmental regulation offers a significant opportunity for opinions to be turned into actions against construction projects.

The client's relationship with the *customer/user* covers knowledge of customer/end-user *needs* (for known and unknown customers/users). In order to meet sufficient requirements, the client should understand customers'/users' *desires, needs* and *economic* circumstances (Johansson and Svedinger, 1997). The client must also consider the customer's/user's *market situation*. There are, nonetheless, limitations in customer orientation. Existing *resources, environmental considerations* and *technical conditions* are physical limitations on customer orientation (Gerdemark, 2000). There has to be balance between *construction, maintenance* and *economy* [finance] in the decision process for the client (Engwall, 2001).

A commonplace goal in the industry is the integration of actors involved in the construction process. The organizations to which the different actors belong have different personnel and material resources as well as different values and norms. It is the organizations that the actors belong to that create the limits for how they think and engage in the construction process. This factor influences the outcome of the process (Ericsson and Johansson, 1994). Members of a temporary organization tend to found their actions on their mother organization's beliefs (Josephson, 1994). Clients must also handle the legal relationships with these actors (Johansson and Svedinger, 1997).

The briefing process

The context for the decision-making in the early stages of the construction process is briefing, the aim of which is to define the operational demands and to support the development of the business process. Decision-making should be seen as a rational process where the decision-maker has all the information needed. Rational decision-making therefore presupposes that there are no uncertainties and indistinctness in the process. Uncertainty and indistinctness in decision-making situations impose limitations on its rationality – so-called bounded rationality (Simon, 1957; March, 1994). According to March (1994), decision-makers normally take few alternatives into consideration and look at them sequentially rather than simultaneously. This decision process comprises two important dimensions: the apprehension between cause and effect, and preferences about conceivable outcomes (Table 2.1).

Table 2.1 Types of decision problem

		Preferences about conceivable outcomes	
		Certainty	Uncertainty
Apprehension between cause and effect	Certainty	Well-structured decisions	Compromise decision
	Uncertainty	Appraisal decision	Ill-structured decisions

Source: adapted from Thompson (1967) and Simon (1957).

The difference between well-structured and ill-structured decisions is that well-structured decisions are repetitive and perfunctory. Ill-structured decisions are, on the other hand, unique and demand more scrutiny of the mapping of alternative and influencing factors (Gareth, 1995; Thompson, 1967; Simon, 1957). Well-structured decisions can be based on fact and can be proven to be true or false. Ill-structured decisions, however, can only declare a sought-after situation and lead to the selection of a final objective. According to Ekman (1970), the well-structured decision can be characterized as an action-decision and the ill-structured decision as a goal-decision.

The briefing process is divided into two stages according to Ryd (2001): strategic briefing and operational briefing. Strategic briefing concerns the business related issues, i.e. the core business, whereas operational briefing relates to delivering the 'technical project' (Yu *et al.*, 2006). The focus is, in many cases, on operational briefing, because the information processed there is more familiar to construction professionals and so they can rely on their experience (Barrett, 1999).

The process of capturing operational needs is divided into four steps: goal formulation, operational needs, analysis of operational needs and, finally, formulation of the products portrayal (Fristedt and Ryd, 2003). The initiation of the briefing process concerns the strategic issues of the organizations and is often neglected. Depending on whether the client organization is unitary or pluralistic, the briefing process must underpin more or less sophisticated processes to create common understanding of the problem to be solved (Green, 1996). The briefing process should be seen as a social process based on iteration and learning instead of a straightforward technical matter of problem solving. According to Katz and Kahn (1978), a problem can be solved by previous knowledge of a similar problem, but if the problem instead has the characteristic of a dilemma, where all alternatives have a negative outcome, there needs to be a new formulation of the problem.

In order to handle insecurity, organizations usually create an administrative structure where the problem space divides into different areas of decision-making responsibility (Axelsson, 1981). The need for information and structure increases with higher grades of non-programming in the nature of the decision (March and Simon, 1993). Organizations can also have interest in the environment that affects the organization's decision by exposing the organization to information (Katz and Kahn, 1978). The information that the organization collects is usually used to confirm decisions already taken, instead of providing a base for decisions to be made (March, 1998). Decision-makers search for information, but they often see what they expect to see. Individuals are especially prone to forming expectations about the future, based on known empirical connection and information about the present situation (Simon, 1957). Knowledge acquired by the client in the early stages of the construction process is more tacit than explicit and needs systems to handle it (Winch, 2002).

Role of the change agent

Managing change is important, but to lead change is crucial (Kotter, 1995). The action needed to alter behavior requires leadership. A change process needs a clear vision and strategy that is based on a thorough examination of the present situation. According to Kotter (1995), there is a difference between management and leadership – see Table 2.2.

Management creates a certain degree of predictability and order, but leadership creates an environment that supports and facilitates change (Kotter, 1995). Katz and Kahn (1978) define leadership as 'any act of influence on a matter of organizational relevance'. The essence of leadership has to do with influential increments which go beyond routines. Leadership taps bases of power, such as reference power and expert power.[1]

In order for a specific change to occur, the new idea must diffuse to the intended recipients and they have to adopt it. The social and communication structure of a system facilitates or impedes the diffusion of change in that system. Change agents usually introduce change into a recipient system. The success of the effort to secure the adoption of a wanted change is related to the change agent's effort in contacting the recipients (Rogers, 2003).

Research project

This research project underpinning this chapter deals with the transformation of requirements in the early stages of the construction process. At this point, clients must define the product. At the same time, clients need to consider all stakeholder expectations. This part of the construction process is also crucial if clients want to act as change agents, since most of the strategic choices are made at this point. The main question to be answered is, therefore, about how construction clients can contribute to change in the construction process.

Research methodology

The empirical evidence in this study was collected through interviews. The interview approach was selected because the researcher was interested not only in *how* decisions were made (normative decisions), but also *why* and

Table 2.2 The difference between management and leadership

Management	Leadership
Planning and budgeting	Establishing direction
Organizing and staffing	Aligning people
Controlling and problem solving	Motivating and inspiring

what (descriptive decisions). The empirical material in the main study was collected through interviews with six clients within different organizations (Table 2.3). The criteria for the selection of respondents were that they were all professional clients in the sense of being almost continuously involved in the construction process.

The interviews were conducted in three stages, with both quantitative and qualitative data (see Table 2.4).

During the interviews, respondents graded the factors from ten = most important to one = least important. The aspect of *construction industry* was not graded by the respondents. This was covered by qualitative questions and is presented below. For each of the factors within the aspects of owner, society and customer, there were several keywords representing each factor. In total, the questionnaire had 20 keywords for the respondents to grade.

In order to complete the decision process, a pre-study was undertaken where interviews were conducted with 12 consultants (architects, technical consultants and project managers) normally involved in the early stages of the construction process. They were selected from both large and small companies and had a local and national presence.

Table 2.3 Interview sample

Type of company	Geographical distribution	Respondent
Private	Local	Owner
Private	Local	Manager
Private	Regional	Manager
Publicly owned	Local	Manager
Publicly owned	Local	Manager
Public organization	Regional	Manager

Note
The interviews were conducted in three stages, with both quantitative and qualitative data (see Table 2.4).

Table 2.4 Interview data gathering

Stages	Quantitative data	Qualitative data
Interview 1	Grading of most influential factor	How the graded factors influenced the client's decision
Interview 2	Grading of factors first examined	How the graded factors influenced the client's decision
Interview 3		Basic data for decision-making regarding the six most important factors

Results from the interviews

The result of the pre-study showed that clients had difficulties in formulating their demands and criteria for the outcome of the process. This can indicate that clients do not have a good basis for decisions in the early stages of the construction process. The consultants also felt that the clients had more or less transferred the client's role to external project managers and they (the clients) were not active in the construction process.

The empirical results from the main study are shown in Table 2.5. The graded factors from interviews one and two are weighted together.

The results show that the respondents considered the customer aspect as most influential over decisions in the early stages of the construction process. Second in importance is the owners' perspective and, third, society.

The two main client groups in this study are private and public. Table 2.6 shows the difference in grading between them.

The public organizations placed a slightly higher grading on customer aspect, owner aspect and much higher grading on the society aspect in terms of affecting their decision. To further understand the process and how decisions are made, each of the aspects is presented with the results from the qualitative interviews.

Table 2.5 Graded factors

Owner	Economic demands	7.4	6.8
	Goals	6.1	
Society	Opinion	4.8	
	Laws	5.9	6.1
	Regulation	7.7	
Customer	Solvency	8.6	
	Customers' demands	9.0	8.4
	Market situation	7.5	

Table 2.6a Public organizations

Owner	Economic demands	7.0	7.0
	Goals	7.0	
Society	Opinion	6.4	
	Laws	7.8	7.0
	Regulation	6.8	
Customer	Solvency	8.6	
	Customers' demands	9.9	8.9
	Market situation	8.3	

Table 2.6b Private organizations

Owner	Economic demands	7.8	6.5
	Goals	5.2	
Society	Opinion	5.2	5.7
	Laws	3.2	
	Regulation	8.7	
Customer	Solvency	8.5	7.8
	Customers' demands	8.2	
	Market situation	6.8	

Relation to the owner

In order to confirm the existence of economic demands from the owner, several cost estimates were found to be used during the construction process. At the start, the cost estimate was stated in broad terms, becoming more detailed further into the process. Normally, the cost estimate was geared to requirements from the owner in terms of interest rates, rates of return and payback periods. There were different approaches to how they handled cost estimates in the organizations, depending on the uncertainty of the project in question. In some cases, respondents even used two separate consultants to estimate the cost in order to get as accurate a result as possible. All respondents agreed upon the difficulty of getting the cost estimate right. This factor had a strong relation to the *market situation*. If the cost estimate was too high (and would result in high rents) the project had to be reconsidered or rejected. This aspect has a connection with customer/user needs too, if the economic demands from the owner do not match conditions in the market in terms of construction cost and achievable rent/returns.

The owners' value was measured in this inquiry by how clear and consistent the organizations' business concepts were. There was a large discrepancy between the different organizations. It was only the private organization with a regional presence that had a clear strategy for its actions. The smaller, local organizations instead saw opportunities and reacted to them. The public organization had a more diffuse business concept (rather a description for action), but it had a higher weighting for the owner's interests in its decision-making.

Relation to the customer

This aspect attracted the highest grading in this study. The public organizations had, however, higher mean values. All respondents regarded customer/user needs as important input in the early stages of the construction process. There was, however, a discrepancy in how they handled the confirmation of customer/user needs in the early stages of the construction process. If they had a known customer/user (i.e. a specific customer) that customer was involved in the process directly. For clients with an unknown customer/user (e.g. rental apartments) they used their experience to decide on the concept.

The relationship with the customer is characterized by how clearly the customer is aware of what is needed. The handling of customer/users needs was problematic according to the respondents. Mostly, they felt that customers/ users had difficulty in expressing their needs and they had problems in understanding the way the construction process was organized and managed.

One question during the interviews covered the concept of customer and the views they and their organization held. Respondents viewed customers as part of two main groups: professional and non-professional. The non-professional customer had to be led through the process by clients asking the right questions in order to define needs. In some cases, customers/users had too high expectations of the construction process and an unrealistic view of the cost required to meet their needs.

Relation to society

The aspect of society represents laws and regulations concerning the construction industry and belongs the variable factor of opinion. The public organizations had a higher grading for this aspect. They related it to their responsibility to society for their organizations. For private organizations, this aspect was not that influential in their decision-making. All respondents said that they always communicated with the authorities regarding building permissions and other matters related to laws and regulations, irrespective of whether or not the project needed permission. They were all very keen that their projects were correctly established.

There is a connection between the aspects of society, owner and customer. The owner's interests and customer's/user's needs can sometimes prove to be the opposite of what the regulations require for buildings.

Relation to the construction industry

This aspect deals with the integration of all actors involved in the construction process. Throughout the interviews, respondents were asked which actors were involved in the process and what their contribution was. Normally, there were architects involved in the early stages of the process and, in some cases, other consultants dealt with the cost estimate. The view from the respondents for the reason to bring in other competences was mostly a question of time: they did not have the time to do the work themselves. Mostly, the clients used their own organization to process the necessary basic data for decision-making. It was mainly in technical areas where they used external sources of competence.

Respondents discussed the market situation and the economic risk for the customer's/user's business. Respondents revealed that there was often a tension between the customer's/user's needs and the cost of the project. Construction cost which leads to rent levels that do not match the customer's/user's means or the owner's economic demands could frustrate the entire process.

The empirical evidence shows that four aspects (customer, society, owner and construction industry) have different impacts on clients' decisions in the early stages of the construction process (Table 2.7).

The aspect of construction industry relates to the coordination and use of the many actors in the construction process. The view from respondents of the kind of competence actors provide to the process differed. Mostly, they revealed that they used external sources of competence to compensate for lack of time in their own organization when issues arose. In order to increase the benefit of external actors' competence, clients must engage with them earlier in the decision-making process. This also demands a clear vision and goal for the decision process in order to capture the tacit information of the different aspects (i.e. owner, customer, society and construction industry). Despite the two distinct types of construction client in this study (private and public), there were similarities regarding the decision-making process. Indeed, the three different aspects were graded in the same order. The difference was in relation to society. There, public clients had a higher grading, which could indicate that their decision-making process is more pluralistic.

Table 2.7 Impacts on clients' decision-making

Aspect	Impact on decision	Tension	Obstacle for change
Owner	Medium graded, large impact	Unclear value from owner Public organizations higher grading, but diffuse demands from the owner Cost as evidence base	Focus on cost Low impact on goal, mission and value
Society	Low graded, high impact	Clear demands, more tension for public client (opinion)	Low impact on change
Customer/ user	High graded, large impact	Discrepancy in evidence base Problem in formulating needs (inexperienced customer) Experience from client as evidence base	Focus on cost Low impact on integration of demands
Construction industry	(No grading)	Low participation; low faith in knowledge contribution	Focus on controlling and problem solving

Conclusions

The purpose of this chapter was to investigate how construction clients handle the transformation of requirements in the early stages of the construction process, in order to contribute to change in that process. The tensions between the different and potentially competing aspects during the briefing stage indicate that the process has the characteristic of ill-structured decision-making. The approach adopted by clients was, however, to treat the decision-making process as if it were more structured. The lack of systematic enquiry in the briefing stage decreases the opportunity for clients to initiate a change process. The change process needs a clear objective and vision that is shared between the participants in the process. The lack of involvement of other participants is a barrier that clients should consider if they are to act as change agents.

The main question to be answered in this study was how construction clients can contribute to change in the construction process. Clients' decision-making processes in this study were concerned more with management (planning, budgeting, controlling and problem solving) of the process than leadership. Respondents were mostly concerned about how to manage the different factors. Frequently, they referred to basic data for decisions that were linked to operational briefing. The strategic issues for the construction process were not taken into consideration to the same extent. In order to act as a change agent, clients need to become an integrator of requirements from the various stakeholders (owner, customer, society and construction industry).

Note

1 According to Katz and Kahn (1978) reference power depends on personal liking between leader and follower. Expert power depends on the knowledge and ability of the leader.

3 Stakeholder engagement in real estate development

Stefan Olander, Rasmus Johansson and Björn Niklasson

Introduction

In the early stages of a real estate development project there is uncertainty as to how the project will be affected by external factors such as stakeholders. In the planning and development process there is also the chance that stakeholders will affect project outcomes, which increases the risk of additional costs impacting on the project. There is, therefore, a need for the developer to understand the process in order to determine the total cost of a development with some degree of certainty.

The research reported in this chapter aims at identifying costs that arise from the impact of external factors in the planning and development process leading up to final approval by the local authority. The research focuses on direct costs resulting in real payments; even so, indirect costs in the form of internal rates of return and delayed revenues are also discussed. This chapter discusses two approaches to the engagement of external stakeholders and from them offers an appropriate strategy for development projects in the future.

Literature review

Stakeholder theory

Freeman (1984) describes the concept of stakeholders as any group or individual who can affect, or is affected by, the achievement of a corporation's purpose. This definition is a development of an earlier stakeholder definition that Freeman traced back to a Stanford Research Institute memorandum in 1963. The memorandum states that stakeholders are those groups without whose support the organization would cease to exist. As a consequence, Freeman (1984) states that the stakeholder approach is about groups and individuals who can affect the organization and the behavior of managers in response to those groups and individuals. Phillips (2003) adds that stakeholder theory should be concerned with who has input in decision-making as well as who benefits from the outcomes of decisions.

Thus, for real estate development projects, it would be the responsibility of the project manager to respond to the needs and expectations addressed by the project's stakeholders and to be concerned with how decision-making is exercised.

There has been debate on defining stakeholders, with Freeman's (1984) definition now being regarded as rather broad, because it merits all to be stakeholders. If everyone is a stakeholder of everyone else there is little value-added from the use of the stakeholder concept (Phillips, 2003; Sternberg, 1997; Mitchell *et al.*, 1997). The view expressed in the Stanford definition – those without whose support the organization would cease to exist – is regarded as narrow since relevant groups would be excluded. Post *et al.* (2002) state that the fundamental idea is to have a stake in the organization, and they define stakeholders as those who contribute voluntarily or involuntarily to the organization's wealth-creating capacity and activities: they are, therefore, its potential beneficiaries and/or risk bearers. Donaldson and Preston (1995) identify stakeholders through the potential harms and benefits that they experience or anticipate experiencing as a result of the organization's actions or inactions. In the early stages of a real estate development project the most influential stakeholders, apart from the developer, are the local authority and those members of the public who consider themselves to be affected by the development (Olander and Landin, 2005; Olander, 2007).

Planning process and development agreements

The planning process is about setting frameworks and principles in order to guide the location of development and infrastructure (Healey *et al.*, 1999). It involves rules and regulations giving certain groups or individuals the right to use land and provides authorities with the means to exert their influence on land use (Larsson, 1997). The planning process includes local and national policies, rules and regulations as well as planning traditions. The key resource in a real estate development process is land and the central role is exercised by the landowner (Verhage, 2002; Barker, 2003). There is a delicate balance between public activities, for instance, the control of land use and the right of landowners to develop land according to their own wishes (Larsson, 1997).

In the Swedish planning process, the local authorities have, by law, significant control over land development within their area. It is the authority that decides on an exclusive basis which real estate developments to approve and can, with a development agreement, control the outcome of such developments. Additionally, vague rules and legislation concerning the planning process add to the uncertainty felt by real estate developers (Riksdagens revisorer, 2001; Olander and Landin, 2008). The risk is that the local authority uses its advantage in the process to force the developer into an unbalanced development agreement (Kalbro, 2002) that adds cost to the real estate development.

The development agreement is the civic tool for controlling land use within designated areas. It regulates issues concerning the purchase of land (if the authority is the present landowner), type of facilities to be developed, timeframes, responsibilities for public streets and green spaces. The development agreement is made within the boundaries of civil law (Miller, 1993; Sohtell and Sundell, 1993). Thus, the development agreement is basically dependent upon the respective bargaining positions of the developer and the local authority. At the signing of a development agreement there can be two situations: one is that the authority is in need of a particular development and may be willing to take a larger economic responsibility for it. In cases where the developer is initiating the development, the authority may consider itself to have a stronger position and may place tougher economic responsibilities on the developer (Wedegren, 1997).

Barker (2004) argues that a more effective planning process would be characterized by a decision-making process that considers the wider costs and benefits of the proposed development in response to market signals. The importance of information and communication in the planning process and how the different intervening stakeholders (e.g. real estate developers, planners, local authorities and the public) may use or perceive it should not be neglected (Silva, 2002). The extent to which developers perceive different aspects of the planning process, rightly or wrongly, may have an effect on their decision to engage in new real estate developments (Olander and Landin, 2008; Olander, 2006).

Perspectives on the planning process

Planning policy can be interpreted and implemented differently in response to local circumstances (Midgley, 2000). Planning authorities and their policies set the framework within which a development can take place (Verhage, 2002) within the limits of national laws concerning the development of the built environment. At the local level, the planning process may adopt different approaches to new development. Some authorities wish to limit new development, while others may welcome almost any development (Monk and Whitehead, 1999).

Designated land use through the planning process is one constraint, among others, on land supply (Barker, 2003). The planning process is complex and timescales are often seen as unacceptably long, with the requirements of planning used to prevent development (Barker, 2004). This process influences the amount of land that is made available and is one reason for high development costs. The planning process has reduced its focus on integrated planning, concentrating instead on the promotion and development of separate projects (Healey *et al.*, 1999; Khakee and Barbanente, 2003) which forces housing developers to have a more or less developed project with resources already bound in land acquisition and design before the planning process can begin.

Developers face a range of significant market and planning risks. These result in a sector that is reluctant to invest over the longer term (Barker, 2004). The level of uncertainty in the planning process also increases the level of risk for developers (Barlow, 1993). This uncertainty can act as a constraint for new organizations wishing to engage in real estate development, because they do not have the financial strength to endure a long and uncertain planning process. The active role of stakeholders in the planning process can, in some situations, become an adversarial affair where conflict arises between different stakeholder interests (Olander and Landin, 2005). This adds to the uncertainty housing developers might already experience from the process.

Other aspects

In recent years, there has been a debate in Sweden about how the planning process affects the real estate development process (Henecke, 2006; Henecke and Olander, 2003; Boverket, 2002; Riksdagens revisorer, 2001). When land supply is locally regulated, lobbying for and against a development becomes more extreme and easier to organize (Barlow, 1993). Furthermore, the local political framework is more prone to pressure from special interest groups. In Sweden, the potential resistance from locally-based special interests has been highlighted as one problem facing the planning process for new developments. Yet, there is a view that the handling of locally-based special interests is an integral part of a democratic planning process. Thus, there is a balance between a time and cost efficient land development process and a democratic process of influence from locally-affected stakeholders (Henecke, 2006) that needs to be addressed by planners as well as developers.

Case study

Description and objectives

The research is based on a case study consisting of two real estate development projects in south-west Sweden. In each project, documents have been examined and stakeholders have been interviewed. The study is limited to direct costs visible in documents and to the early stages of the development process up to the final approval of the project by the local authority. The length of this process has also been examined, enabling a discussion on the indirect costs that internal rates of return and delayed revenues would bring.

Research methodology

The purpose of the research is to map the type of effects, and their associated costs, found in the real estate development process. The case study approach was chosen as the primary research method in order to obtain in-depth knowledge of external impacts on the process. The purpose was not to create a model for analyzing external impacts, but rather to gain an understanding of them in a way that could eventually be utilized in further studies.

A questionnaire survey could have been an alternative method for obtaining the relevant information. The benefits of choosing this method would have been the opportunity to obtain information and knowledge from a large number of real estate developers and projects. The chosen research method (i.e. case study) describes the qualitative phenomenon of external impacts in the real estate development process. The results and analysis of the research presented here could, however, be a baseline for the design of a questionnaire for a more comprehensive study of impacts and how real estate developers react to them.

A case study is an empirical inquiry that investigates a contemporary phenomenon within its real-life context, especially when the boundaries between phenomenon and context are not clearly evident (Yin, 2003). Case studies can be either quantitative or qualitative; in the present research, a qualitative approach was chosen. A qualitative case study focuses on matters of insight, discovery and interpretation rather than testing a hypothesis. A qualitative case study can be defined as an intensive analysis of a single phenomenon, at the same time as the whole is under scrutiny (Merriam, 1998).

Both projects are located in expanding communities. The projects, Västra Hamnporten (West Harbor Gateway) in Malmö and Lomma Hamn (Lomma Harbor) are relatively similar considering their size and location, as well as being close to the sea. They are considered attractive for both residential and commercial uses. A criterion of choice was that final approval of the chosen projects was not more than two years in order to ensure that the information gathered from documents and interviews was reasonably up-to-date.

The main similarity in the projects, described below, is that they are large developments in centrally-located harbor areas. Both areas are no longer used for their original purpose; instead, they are being developed into residential and commercial areas where the projects are part of a larger development. There are, however, some differences. The local authorities of Malmö and Lomma are different in size and, therefore, do not have the same organizational resources. On the other hand, the smaller authority, Lomma, does not have as many development projects as Malmö, which enables it to focus more on the ones it has. The respective real estate development companies, JM and Midroc Property Development, are also different in terms of size, resources and number of ongoing projects.

Project 1 (*Lomma Harbor*)

Lomma is a medium sized community of approximately 18,000 inhabit-ants with good communications to the main cities in the region, mainly

Lund and Malmö. Lomma Harbor is an entirely new town district and will increase the population of Lomma by 3,000. The development consists of housing and commercial and public buildings that require completely new infrastructure in terms of roads, water supply and sewage system. The land was mainly used for industrial purposes. Prior to 1977, a factory produced asbestos cement products on the site, but closed after health risks were identified.

The part of the development that is the subject of this study is owned by JM. In addition, two development areas are owned by other developers. The real estate studied here consists of about 43,000 square meters (gross floor area) and 330 dwellings, with an opportunity to combine them with commercial areas. The following specific conditions for the project were identified.

- A project team was assigned by the local authority with the aim of coordinating any concerns of the developer with those of the authority.
- JM took the entire financial risk of providing infrastructure for the new town district.
- There were delays in the project due to appeals against the local authority's decision to approve the development.

Project 2 (West Harbor Gateway, Malmö)

Malmö is the third largest local authority in Sweden with nearly 300,000 inhabitants. The development studied here is a part of the development area that consists of West Harbor Gateway, Dockan and Turning Torso (a landmark, high-rise building). The entire area was formerly used for industrial purposes, and the proposed development consists of dwellings, hotels and offices totaling 42,300 square meters (gross floor area). The real estate is owned by the developer, Midroc Property Development, and the following specific conditions for the project were identified:

- There had been many changes in the proposed development due to additional evaluations and shifts in market conditions.
- The development would occupy a key position in an expanding area, thus contributing to the local authority's interest in its size, shape and design.
- The development is located at the main entrance to the area of West Harbor Gateway and Dockan, and has resulted in increased costs for infrastructure.

Data gathering

The empirical data consist of project documentation and interviews. The former were mainly invoices covering costs that had adversely affected the project budget and public documents concerning planning and development agreements. Interviewees were conducted with the project developers, project managers and local authority planning officials. A total of nine semi-structured interviews were undertaken covering subjects such as project implementation and perceptions of the public planning and development process. Analysis and evaluation of the empirical data have provided an in-depth insight into the projects and external impacts upon them.

Research results

External impact can be divided into direct and indirect. The direct impact is mainly those commitments that the real estate developer agrees to make when signing the development agreement. In the two projects, direct impact can be described as:

- Transfer of land within and outside the proposed development from the real estate developer to the local authority without monetary compensation.
- Responsibility for decontamination and preparation of the land and adjacent properties.
- The construction of roads and other infrastructure outside the proposed development.
- Parts of the proposed development are to be reserved for the purpose of social services (e.g. childcare and schools) without monetary compensation.

In addition, costs have arisen in the planning process from further investigations and evaluations demanded by the local authority.

The total direct costs related to external impacts for Project 1 were SEK22 million[1] and SEK14 million for Project 2. Both projects have planned gross floor areas in the region of 43,000 square meters. Some of the costs for Project 1 arose from a more thorough analysis early in the planning process. In both projects, the developers have been willing to accept the costs of providing public facilities because of promising forecasts of future revenue.

The direct impact in relation to the total investment for these projects is about 2 percent, which in one sense could be seen as reasonable. If the costs are instead related to the purchase price for the land then it amounts to 25 percent. This indicates that a real estate developer would need to cover substantial additional costs before acquired land could be developed.

Another important aspect is the impact on indirect costs arising from time taken in the planning process. For both projects, there have been substantial, additional evaluations, which have prolonged the planning process. The developers assess, in addition, that the planning process has been delayed by one year due to ineffective local authority management. For both projects the total time of the planning process amounts to four-and-a-half years compared with a normal period of 18 months to three-and-a-half years: this will cause indirect costs.

If a reasonable internal rate of return is assumed to be 7 percent, the indirect cost due to the binding of capital in the purchase of land will amount to approximately SEK25 million for Project 1 and SEK18 million for Project 2. If the delay in the planning process in relation to a normal period is assumed to be two years and the revenue is assumed to be SEK1,000 per square meter (gross floor area), the indirect costs due to delayed revenues will be, for both projects, SEK8.6 million. These figures should be viewed as examples of possible indirect costs based on assumptions of future revenues and rates of return. If this indirect impact were added to direct impact, the total external impact would amount to 50–60 percent of the purchase price for the land. This indicates that the hidden cost of binding capital is the larger part of the external impact and due mainly to a prolonged planning process.

This study should be seen as a baseline for future studies. It shows that the external impact on a new real estate development will mean substantial costs, both direct and indirect. This will be an obstacle for small companies wishing to engage in real estate development and can contribute to weaker competitiveness, because of the substantial organizational and financial resources needed to manage the uncertainty as well as the costs that arise in the process.

Conclusions

The external development costs identified in the study are not entirely unreasonable for the developer to accept. Some costs are necessary for the accomplishment of the project and for assuring acceptable quality in the development. It is the uncertainty of external impacts, and their costs, occurring during the planning process and arising from development agreements that constitute the major barriers to land acquisition. Clearer regulations and policies are necessary in order to estimate the duration of the planning process and the costs likely to be incurred over that time.

The local authority, as the representative of the community, has an interest in the quality of real estate development. The management of the planning process can be improved by streamlining the early planning stages through identifying the key issues. Improvement here should give both the authority and the developer common ground for future decision-making in the development process. These actions should decrease the uncertainty related to external impacts on the planning process.

The statutory position of the authority means that it is a strong player in the planning process with the power to decide what is reasonable in each project. A signed development agreement is a condition for final approval of a proposed development, which often makes the developer the weaker partner when negotiating the terms of the development. The developer needs to have a clear understanding of the influence that various stakeholders will have on the project. As one of the primary stakeholders, the authority has the opportunity to become a positive force in real estate development rather than a constraint.

Note

1 Twenty-two million Swedish crowns.

4 A systems approach to construction supply chain integration

Ruben Vrijhoef and Hennes de Ridder

Introduction

In construction, the production system and supply chain in particular have been deemed to be relatively disintegrated. A more integrated approach to construction has been often put forward as a solution to the many problems and deficiencies existing in it. At the same time, restrictions imposed by integration in construction have been recognized, because of its temporary and complex nature. Here, the approach is to regard a construction supply chain as a system and to apply systems engineering to increase the coherence of the supply system. The underlying principle is that a production system such as a supply chain delivering a single product should not be fragmented, nor should it consist of distributed functions. Instead, supply chain integration must lead to improvement by developing a more stable, repetitive production environment, similar to what is common in other industries. The premise is that the construction supply chain would function better when approached and reconstituted as a single entity in the form of an extended enterprise. In a way, the broader issue is whether construction could or should develop itself toward the standards and practices of a conventional, more integrated, supply-driven industry. This chapter provides an overview of research that is applying a systems approach to the creation of a model for supply chain integration in construction. It offers insights into a number of the *building blocks* found in theory and practice, which can be used in the model building process.

Theoretical building blocks

Viewing the supply chain as a system

Systems theory views the world in terms of collections of resources and processes that exist to meet subordinate goals. Two aspects of systems theory are of particular importance for supply chains: synergy and entropy. *Synergy* means that the parts of a system working together can achieve more than the sum of achievements that each one would achieve separately. *Entropy* refers to the necessity of feedback across the chain to prevent debilitation of the system (New and Westbrook, 2004). Hassan (2006) has suggested the application of system engineering to the design and formation of supply chains. The structurist character of systems thinking can be helpful in building the structure and operations of the supply chain in a systematic manner, assuring its effective functioning.

In terms of systems typology, supply chains are human activity systems and social systems, consisting of actions performed by individuals and groups of individuals, e.g. firms (Checkland, 1981). Supply chains can be characterized as networks between economic actors (i.e. firms), engaged in a voluntary relationship to produce and deliver a product or service. Rouse (2005) considers the nature of firms as systems and supply chains as *systems of systems*. This is essential to a proper understanding and ability to find integrated solutions to improve firms and systems of firms (i.e. supply chains). Rigby *et al.* (2000) underline the importance of systems thinking for organizational change and improvement, but warn of the risk of underestimating the complexity of reality when translating it into a mental model. Systems approaches are not fully capable of capturing soft factors such as power, trust and human behavior.

Supply chain as a social system

In construction, the relationships between firms are typically maintained for the duration of the project. Supply chains are not merely directed toward minimizing transaction costs, but also toward enhancing the transfer of expertise and systematic feedback on planning, design, construction and maintenance between parties, and ultimately striving for joint value maximization. Increased cooperation and integration between supply chain parties enables delivery of a total product with quality guarantees to the market. Bounded rationality and differences in know-how between firms could be resolved by joint product development. Opportunistic behavior is then replaced by mutual trust, which is necessary for an open dialog and optimal knowledge sharing.

Dubois and Gadde (2002) distinguish between tight couplings in individual projects and loose couplings in the permanent network within the industry, which they term a *loosely coupled* system. The pattern of couplings influences productivity and innovation, and the behavior of firms. In terms of organizational behavior, cultural and human issues such as trust and learning have been shown to have significant implications for construction supply chains (Love *et al.*, 2002). The social systems approach may, therefore, improve not only the performance of supply chains, but also the socio-organizational basis of the inter-firm relationships within the supply chain.

Supply chain as an economic system

In economic terms, a supply chain is a series of economic actors, i.e. firms buying from and selling to each other. From such an economic perspective, the choice of a coordination mechanism or governance structure is made by economizing on the total sum of production and transaction costs (Williamson, 1979). Transaction cost economics (TCE) provides an explanation for the existence and structure of firms and for the nature of coordination within a supply chain (Hobbs, 1996). When transaction costs are low, contracting is used (i.e. market structure), while internalization will prevail for high transaction costs (i.e. hierarchy). Intermediate positions are often referred to as hybrid modes (Williamson, 1991).

TCE recognizes that transactions do not occur without friction. Costs arise from the interaction between and within firms, e.g. information costs, negotiating costs and monitoring costs (i.e. enforcement costs) (Hobbs, 1996). Transaction costs would be zero if humans were honest and possessed unbounded rationality. For a given situation, transaction costs depend on three critical dimensions: asset specificity, uncertainty and frequency (Williamson, 1985a). In addition to these key concepts underpinning TCE (i.e. bounded rationality, opportunism, asset specificity, uncertainty and frequency), Milgrom and Roberts (1992) offer *difficulty of performance measurement* and *connectedness to other transactions*. Both are relevant from a supply chain perspective and may be influential in reducing transaction costs. Improved collaboration and communication in the supply chain will also reduce transaction costs.

Supply chain as a production system

The supply chain is geared to the delivery of a product or service to an end market or single customer. This implies a production process which is purposive. The managers of the production process need to ensure that the purpose of the process is achieved effectively and efficiently by addressing the transformation (i.e. conversion), flow and value aspects of production in an integrated manner (Koskela, 2000). In terms of the firm, both primary and support activities are aimed at the delivery of customer value and, as a result, revenues and profit for the firm (Figure 4.1).

Supply chain as an organizational system

Firms, as well as supply chains, are organizational systems built from various vital elements that enable them to function as they do. By viewing organizations as systems of flows, Mintzberg (1979) identifies various system representations of organizations, together making up the structure and infrastructure of organizations as systems of formal authority, regulated flows (i.e. material and information), informal communication, work

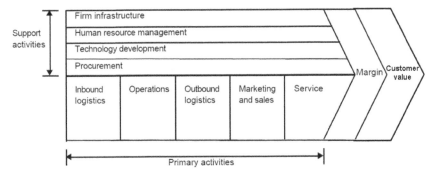

Figure 4.1 Value chain aimed at customer value (source: adapted from Porter, 1985).

constellations and ad hoc decision processes. Typically, the supply chain is a *system of systems* or a *superstructure of organizations*. Firms along the supply chain perform distributed production activities and business functions. This raises the issue of the core competences of firms (Prahalad and Hamel, 1990) together making up an extended enterprise. In construction, this issue relates to the idea of the quasi-firm, a term coined by Eccles (1981b).

Research project

A model of supply chain integration is being created using an organizational systems approach in which the supply chain, as noted above, is viewed as a *system of systems*. First, a generic model is being built by applying theoretical building blocks from the four theoretical perspectives presented above, namely economic, social, organizational and production system. Next, the generic model will be specified and illustrated by adding empirical building blocks from practical examples, i.e. case studies of supply chain integration inside and outside construction.

Fragmented construction supply chain

The construction industry has often been characterized as complex, a reference to the demography of the industry – as exhibited by many SMEs and specialist firms – and the organization of construction activity, including the configuration and coordination of construction supply chains. Indeed, construction is far less structured than other industries, with a vast network of actors of different kinds involved in the development and construction of a built object (Figure 4.2).

The production situation in construction could also be equated to assemble-to-order production and capability oriented production systems (Wortmann, 1992). Alternatively, construction could be observed as a

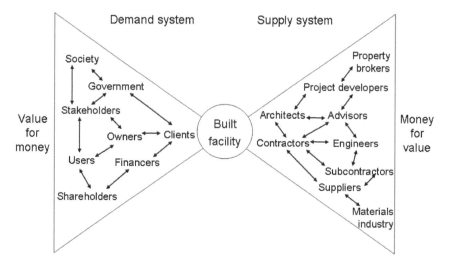

Figure 4.2 Construction demand and supply systems (source: Vrijhoef and de Ridder, 2005).

make-to-order, design-to-order or even concept-to-order production system (Winch, 2003). The fact that construction is often a demand-driven process, and design is often disconnected from production, leads to various problems in production. The producer is not the designer and production is very much influenced by craftsmanship. Moreover, production involves many crafts and these tend to be handled by relatively small firms. This can cause a persistence of problems upstream in the supply chain and, often, worse problems downstream, because of the mechanisms of causality and interdependence at work within the supply chain.

In most construction projects, the end-customer is in at the start as well as the end of the entire process, and therefore exercises a dominant role. This also causes reaction in construction supply chains and hampers pro-activity. This is one reason why construction products are rarely launched and marketed in the same way as in other industries, e.g. consumer goods. Most contractors have no interest in integrated end-products. Most products are not standard and processes are not repetitive, often causing high levels of waste (Vrijhoef and Koskela, 2000).

Integrating the construction supply chain

Low levels of integration and repetitiveness in construction lead to problems and underperformance of the supply chain as a production system (Vrijhoef and Koskela, 2000). An approach to resolving this problem is to apply concepts that increase integration and repetition within and between project supply chains, such as in the case of partnering arrangements (see, for example, Bresnen and Marshall, 2000). Previous work points to the need for more alignment and more structured ways of organizing and managing the construction supply chain. Systems engineering can help in the sense that its goal is supply chain integration, which means engineering problems out of the supply chain, i.e. the production system (Hassan, 2006).

Stevens (1989) stresses the importance and opportunities arising from supply chain integration for companies in reacting to market conditions and reducing costs. In order to comply, virtually all companies and functions in the supply chain should be connected, operating as if they were a *factory without walls*. Fawcett and Magnan (2002) argue that often supply chain integration is not fully implemented by companies in a way that the whole channel from *suppliers' suppliers* to *customers' customers* would be integrated. In many cases, they found it is simply impossible to fully integrate an entire supply chain. This is particularly true of temporary and fairly un-integrated construction supply chains. An alternative solution is to integrate the demand and the supply sides separately. This approach calls for two new central roles in the demand and supply system: the demand system integrator and the supply system integrator (Figure 4.3). For instance, clients or lead consultants, architects or engineers would take up the demand integrator role; contractors or suppliers would take up the supply integrator role (Vrijhoef and de Ridder, 2005).

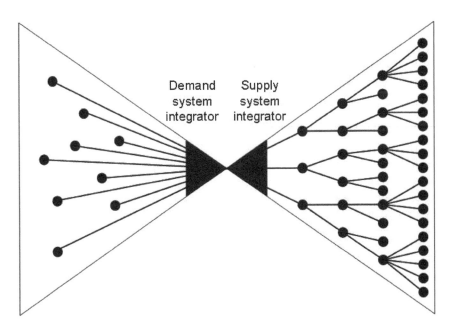

Figure 4.3 The demand and supply system integrator (source: Vrijhoef and de Ridder, 2005).

Theory building through model building

The research reported here adopts the idea of theory building from case studies as introduced by Eisenhardt (1989). The approach is semi-inductive, starting from theory and case studies (building blocks), shaping hypotheses, and from there building a theory (model). This corresponds to the idea of constructive research, which combines analysis of existing phenomena with the building of new concepts. This kind of research is aimed at designing solution-oriented research products, rather than deducing analysis based explanations (van Aken, 2005).

The research approach could be summarized as one of engineering, i.e. engineering a supply chain integration model as if it were a system that should be functional and useful. The engineering process starts by building the generic supply chain integration model using the theoretical building blocks found in the four theoretical perspectives presented earlier. The generic model built from these building blocks will be specified and illustrated by adding empirical building blocks from a select number of case studies of supply chain integration. These studies include descriptive explorations of examples of supply chain integration inside and outside construction. This chapter only allows space for a brief description of four of the cases studies – two cases within construction and two from outside.

The construction case studies are used to describe the supply chain integration strategies applied by different parties, i.e. firms along the construction supply chain including clients, architects, contractors and suppliers. The case studies cover a number of types of construction rather than a single type. In a later phase of the research, guidelines for supply chain integration for different types of construction are derived from the model based on analysis of the case studies and expert opinions.

Case studies: empirical building blocks

Truck manufacturing and shipbuilding

In the early 1990s, the Dutch truck manufacturing industry entered a period of crisis. After drastic measures were applied, most companies recovered and are currently doing reasonably well. One of the measures was to reform and integrate the supply chain. Suppliers have since been integrated into product development, planning and logistics. For clients, the establishment of an integrated dealer network assures direct follow-up of defects and 24-hour roadside recovery and repair.

In shipbuilding, few producers have improved their businesses significantly. Even so, they are globally leading in a few product categories. For these products, they have introduced strict standardization and modularization, and imposed this on their suppliers. This has substantially improved profitability and quality. Some suppliers have effectively become external business units, guaranteeing close links with the shipbuilders.

Housing and commercial building

In the housing sector, a few builders have changed their businesses into suppliers of pre-engineered housing. They offer a choice of houses from a catalog and these can be erected on site within one week. Moreover, their houses can be customized according to a client's individual wishes. Integrated in-house production and pre-assembly assure a smooth process, as well as preventing delays and quality problems. In addition, they arrange for planning and other permissions from the local authority and can also assist with mortgages.

Many Dutch real estate developers have refocused their business on the front end of the supply chain. They have acquired land and existing buildings for development purposes. Additionally, they deliver all services required by their clients including finance, maintenance, facility management and services such as security and catering. Some developers have integrated the supply chain to such an extent that they have become their own clients, in order to find users for their projects once completed.

Comparing supply chain integration

When broadly comparing the examples of supply chain integration inside and outside construction, one can see differences as well as similarities. Differences can be found in terms of pre-engineered products and the integration of the supply chain. Outside construction, the levels of pre-engineering and integration are higher, because levels of repetition are generally higher. Similarities can be found in the mechanisms for integrating design, after-sales and other client-focused services. These mechanisms are common in most manufactured goods sectors.

The characteristics of industries do, however, vary. The production system of each industry has been shaped by its inherent characteristics and history. Production systems in project-based industries such as construction are geared to a product mix that is one-of-a-kind or, at least, one with little repetition. Processes are somewhat jumbled, segments are loosely linked and management challenges are dominated by bidding, delivery, product design flexibility, scheduling, materials handling and shifting bottlenecks (Schmenner, 1993). Fragmentation in the construction industry has long been identified as a major source of complaint, adversely affecting its practices (Turin, 2003). This feature is manifest in the predominant one-off approach that characterizes construction and its *unique-product* production (Drucker, 1963).

Construction can be characterized as a specific kind of project-based industry and has been associated with *concept-to-order* and *engineer-to-order* products to the extent that it is viewed as a type of project-based production, rather than manufacturing. The latter refers to *assemble-to-order*, *make-to-order* or *make-to-stock* types of production system. 'Treating construction as a type of manufacturing obviously neglects the importance of the design aspect of construction, and arguably subordinates value generation to waste reduction, which inverts their proper relationship'; however 'certain aspects of construction could move into the realm of repetitive making' (Ballard, 2005). The general production system types of different industries vary in their focus on either (one-off) designing or (repetitive) making (Figure 4.4).

Implications of supply chain integration

A demand system perspective

Traditionally, clients have played an important and dominant role in construction (Cherns and Bryant, 1984). In terms of supply chain integration, the client's role can be critical because of the initial decision to procure construction works and the way in which procurement takes place (Briscoe *et al.*, 2004). Clients who have the power to shift their procurement strategies vis-à-vis the market are in the position to align the supply chain effectively and implement supply chain integration successfully (Cox and Ireland, 2002). In these cases, procurement strategies must be aimed at establishing long-term relationships in the supply chain. Few advanced and regular clients with buying power have created multi-project environments or manage their procurement through a portfolio approach (see Figure 4.5). If they were to do so, they would increase repetition and create similarities across multiple projects, thus increasing the degree of project certainty and supply chain stability (Blismas *et al.*, 2004). Often these clients have successfully introduced a strategic long-term approach to procurement, which has proven to be particularly effective for certain sectors of construction (Cox and Townsend, 1998).

A supply system perspective

On the supply side, the parties have tended to evolve toward more integrated arrangements through project-independent collaboration with

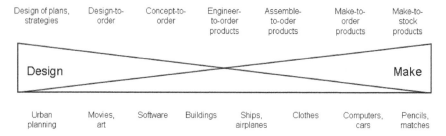

Figure 4.4 Production system types (source: adapted from Ballard, 2005).

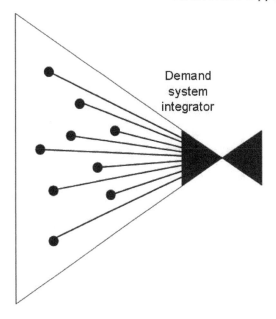

Figure 4.5 The role of the demand system integrator (source: Vrijhoef and de Ridder, 2005).

others in the supply chain and the internalization of neighboring activities or businesses. In both cases, operational and competitive advantages, through higher levels of productivity and efficiency and delivering better client value, are the drivers for this kind of supply chain integration. Normally, this development is led by a focal firm, the system integrator; this could be a main contractor, but also an architect or engineering firm (Figure 4.6).

Conclusions

The use of theory and practical examples from other industries demonstrate the value of supply chain integration in construction. The characteristics of construction demand that any model of supply chain integration for construction must be capable of adaptation according to the needs of individual firms along the supply chain for the different types of construction involved. The systems approach, as proposed in this chapter, is helpful in building a model of integration and for providing a basis for improvement in construction supply chains. This work of model building is an exercise in using theoretical building blocks (concepts) and empirical building blocks (cases) to create a change model for restructuring existing construction supply chains. In order to succeed, all functions along the supply chain need to be decomposed and reconfigured according to their purpose

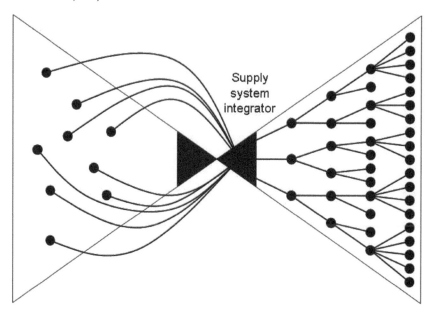

Figure 4.6 The role of the supply system integrator (source: Vrijhoef and de Ridder, 2005).

and the interfaces between them. If this were achieved, endemic problems and irrationalities should be engineered out of the construction supply chain, eliminating many problems and their negative symptoms. A positive side-effect will be that the control of different functions becomes more aligned and centralized, transforming the supply chain into an integrated structure, i.e. achieving the status of the extended enterprise.

Acknowledgements

The authors gratefully acknowledge the financial support of *PSIBouw*, TNO Built Environment and Geosciences, and Delft University of Technology for the doctoral research project covered in the writing of this chapter.

5 Rethinking communication in the construction industry

Örjan Wikforss and Alexander Löfgren

Introduction

Within the framework of a national development program, IT Bygg och Fastighet (2002), a number of pilot studies were carried out on construction project networks, digital document management and cooperation over construction information models. These studies focused on solving the technical problems of information management, although practical experience showed that there is a general resistance toward introduction of the technology within the construction industry. This resistance was not based solely on a perception of technical shortcomings; there were also significant non-technical elements, such as methods and routines, the roles of the various parties involved and the legal and economic prerequisites. Similar experiences have been noted in studies carried out in other countries which have often highlighted the need for improved integration of design and production and cooperation with the client on construction projects. In recent years, the primary causes of the construction industry's poor performance have been identified as its ineffective communication practices, organizational fragmentation and the lack of integration between design and production processes (Dainty *et al.*, 2006).

So far, research and development initiatives in the industry have been dominated by the purely technological development of information and communications technology (ICT). These efforts have not resulted in a comprehensive understanding of how new technology works in project communication if considering human, organizational and process-related factors, as well as those of a purely technological nature. A former chairman of the International Alliance for Interoperability (IAI) has even questioned the work approach of IAI in the development of the international building product model standard, Industry Foundation Classes (IFC). After 11 years of IFC development, its adoption and use in the construction industry is marginal. The ambitious approach of IAI may have focused too much on the model-based world instead of the real world. This has succeeded in IFC being regarded as a theoretical model specification or an academic exercise rather than a useful industry standard for professionals in practice (Kiviniemi, 2006). At the same time, industry has already begun implementing and using new technologies and applications. The large-scale adoption of ICT in construction stands to derive major advantages only if experience of its use can be gained at an early stage.

It is in this context that *project communication* – a newly introduced subject at the Royal Institute of Technology (KTH) in Stockholm – seeks to study ICT in its practical context. Within the realm of this research area, an investigation has been carried out into communication during the design stage for two construction projects: the design work of a new building for the National Defense College in Stockholm and a project run by *AB Storstockholms Lokaltrafik* for the rebuilding of the subway train station at Sockenplan (Wikforss, 2006). The scope of the studies was then expanded to include communication during construction, focusing on mobile work activities on the site and the particular needs of communication (Löfgren, 2006). A comparative study was also undertaken of how four regularly used, Internet project websites function as a means of communication between project participants. This led to a discussion on the need for information technology specifically designed for project management purposes (Wikforss, 2006).

This chapter describes some of the fundamental collaborative communication issues in the planning, design and production phases of construction projects. The chapter introduces the perspectives of project communication research and outlines an initial conceptual framework for developing communication practices combined with supportive ICT as a facilitator for improved organization and management of future construction projects.

Communication in construction planning and design

Case studies

The first case study considered project communication during the final stage of planning for a new building and renovation work at the National Defense College and the Swedish Institute of International Affairs in Stockholm. This is a unique project as, indeed, is almost always the case in architecture and building. A project group is put together for one particular occasion and is disbanded once the work has been completed. During the stage of the design studied here, the need for information exchange between the project participants was intense, time being the critical factor. Demands on participants' performance and accessibility steadily increased, while the amount of information they were expected to handle was extremely large. The focus was on detailed control and coordination of various partial solutions to the whole. Time deadlines, financial pressures and shortcomings in previously developed technical documentation affected relationships and cooperation between the members of the project team. In this environment, the project participants tended to become less careful of how they passed on their information. Communication more frequently took place via informal, direct channels rather than via those originally planned, which were based on the storage of project data at a shared site common to the project as a whole. As different participants focused on different areas and had different, sometimes conflicting, interests the distribution of information and cooperation within the project was adversely affected.

The second case study involved the rebuilding of the Sockenplan subway train station. The construction project can be described as a component of a continuously running system of regular renovation of stations. The participants in this particular project had thus worked together previously and a functional organization for the project management was already available – the ideal conditions, in other words, for ensuring that the project and its communication needs could be managed carefully and in good order. In the case study, it was observed how the project was initiated and planned, and how communication was handled during the planning stage. Initially, the project managers expressed an ambition to organize and control communication within the project via an Internet-based project management network. The project managers set up the network, introduced it to the other project participants and encouraged its use, although they failed to do so on time. Once the network had been set up, even the project managers did not use it to its fullest extent. Documents were not distributed via the project network, but were instead distributed directly among the parties involved, either by email or as normal paper copies. The result was that the use of the project network remained limited. Instead, information flowed in an uncontrolled manner among the members of the project team.

Implications of the findings

The findings from these studies highlighted two different perspectives that are diametrically opposed. The first perspective is that of the project manager. This is the image of the ideal process as it is described in industry-wide documents, contracts, instructions and manuals of various types. It is the image of the orderly process that proceeds in discrete steps clearly defined in advance, traveling along well-signposted information highways. It is an image of the process that is seldom questioned: it constitutes accepted practice. However, project managers have found it difficult to get their teams to adhere to this paradigm. In practice, project participants actually oppose and even obstruct the use of central project sites that project managers wish to use for the exchange of information.

The second perspective is that of the planner and designer. This is the image of the design work that will actually be carried out. It is about issues that are important and difficult to tackle, and about how ideals, facts and value judgments become inextricably mixed in informal, but authoritative, design decisions taken at intervals between the occasions when formal decisions are made. Judgment-based decision making, planning, improvisation and reflection-in-action are key concepts. This paints a picture of a somewhat chaotic work process in which informal contact channels (i.e. shortcuts) and verbal agreements determine the results that will be achieved. It is a picture of a process that can seldom be discussed openly during the project since it is not actually accepted.

In this context, it is interesting to note that ICT is also used for a significant proportion of the informal communication, although it differs from the ICT offered at the central project site. Here, the emphasis is on direct contact and speed of communication. There is, however, a significant risk of losing sight of the bigger picture and the control of the construction project as a whole. Who, for example, will join up the design process with the preparations needed for production to ensure that the proposed building is actually erected? Who ensures that the project team maintains a shared understanding of the project's ultimate objective right up until the time when the building is finally handed over to the client? What is more, the various specialists involved in such projects all use their own jargon, a kind of professional language that keeps others out and maintains the pecking order between the various groups. Meanwhile, the traditional distribution of roles is controlled by stereotypical notions of what others can do and cannot do, and intentional misunderstandings are part of a technique designed to strengthen one's own role and protect one's own personal space in this ongoing game.

In the constant negotiation between the members of the project team, as to exactly where one's duties lie and who is expected to do what, the winner will be the individual who enjoys the advantage of information. For the individual player, the smartest strategy may well be not to communicate everything, not to have heard some piece of information, even to have suffered a slight misunderstanding. This, indeed, may be the real reason why participants are reluctant to publish their information on the common project site. There are perfectly rational reasons for not making a technical solution available to the project network too early on – who wants to risk being held responsible for having spread inaccurate information? Likewise, there are perfectly rational reasons for instead getting in touch directly with a project member you know you can trust, someone you can rely on not to look for faults and demand damages. Project networks are thought up for an ideal situation in which accurate information is exchanged in predictable patterns drawn up in advance. Yet, the conditions under which real projects must operate are typically unclear and unpredictable, and technical solutions remain imperfect for a long time. Professional skills consist of an ability to manage this ongoing search for the end solution, which is why professionals will wait as long as they can before publishing their information.

Construction projects are assembled by gathering different professions and areas of expertise under one 'flag' (Söderholm in Wikforss, 2006).

Typical of such assemblies is that each professional group also bears a set of principles, rules, knowledge domains and professional skills formulated in a certain manner. At the same time as this helps make the profession strong and successful, it also illustrates why they cannot cooperate with other professions particularly well. Taking this professional barrier as the starting point, a construction project can be described as a 'battle of the giants' in which each of the professions involved is fighting for supremacy over the others. But the battle is not fought within individual fields of knowledge. Design engineers and other technical consultants know that the design is the responsibility of the architect, and although they may have their views on the subject, the architect's monopoly of knowledge in this respect is not seriously challenged. When it comes to organizational tools, duties or constellation forms, of which none of the established professional groups holds a previous monopoly, the battle suddenly becomes important. It is not always a battle for the best solution, but rather a contest to establish whose opinions carry the greatest weight and what sort of information is actually of importance.

Communication tools introduced with a purpose of imposing better control and coordination of construction projects are an arena for such knowledge contests. Communication solutions aim at breaking down barriers that professional groups have carefully and successfully built up over a long period. They aim at making construction knowledge more general, thereby challenging the expertise that for decades has become more and more the province of specific professions and home to an ever increasing array of professional jargon. These tools also aim at coordinating activities between professional groups, which today all apply their own special routines and have their own particular ideas as to how coordination should be achieved. This may result in communication tools that are so generally conceived, so shallow and so uninteresting that they can be generally accepted but are hardly ever used; or, someone may take control of the tools and modify them to suit their own special needs, thus obtaining a toolkit that is both sophisticated and functional – at least for a few (Söderholm in Wikforss, 2006).

Communication in building production

In an introductory investigation of problems ahead of an attempt to introduce mobile ICT support at construction sites, work on a construction site north of Stockholm was studied for half a year on a regular basis through direct observations, interviews and document analyses (Löfgren, 2006).

The production environment of the construction site involves a very tight time schedule with full attention to planning, coordination and completion of building activities. Production managers, construction supervisors and superintendents are needed on site to coordinate work, make inspections, conduct environment and safety rounds, and document and follow up ongoing and completed construction activities. The very same people also need to be located at their computers inside the site office ordering equipment and building materials, exchanging digital drawings between architects and design engineers, emailing subcontractors about upcoming work, following up budget figures and invoices as well as preparing deviation reports on construction work with unsatisfactory results. In addition to these are daily production meetings that afterwards need to be transcribed in computer documents and emailed to all involved parties.

Construction projects of today are dependent upon reliable and updated information through a number of ICT-based business systems, communication tools and shared storage servers. Resolving problems arising on-site and critical construction issues requires quick access to necessary information. In order to solve a site problem, production management personnel have to run back and forth between the construction workplace and their computers inside the site office. This leads to inefficient use of managerial resources because production management personnel are occupied at their computers for a large part of their working day. Production managers and construction supervisors often find that they have to be in two places at the same time; at the site office doing administrative work on their computer, as well as being out on the site coordinating work (Löfgren, 2006). Documentation of building activities, production meetings and various inspections often has to be carried out twice: once when it is actually occurring and then again in a computer-based document using different templates.

Even though the purpose of ICT-based business support systems is to improve project communication, their use has led to production managers, construction supervisors and superintendents believing that they are doing the wrong things. For example, whole days are sometimes spent in front of the computer writing up minutes from previous meetings. This has resulted in a negative impact on management presence and leadership in the production site environment. Most of the available project-oriented ICT tools are meant for formalized office use. These tools offer only modest support to craftsman-like construction activities and the unpredictable and mobile environment in which site personnel work. Improving information and communication support for core activities on construction sites has become a strategic challenge for the construction industry in order to increase efficiency and productivity in the construction process (Samuelson, 2003).

Project communication

The dilemma of project management

Both planning and production share a need for rapid access to information and communication in real time. An interesting study object is therefore the communication toolkit commonly used for ICT-based project communication today – the project network. Four different project networks were compared in terms of their basic structure and functions and methods of use (Löfgren in Wikforss, 2006). The aim was to identify the potential offered by each of the networks for coordinating communications within a project and to compare this with how the networks were actually used. The results were based on a large number of interviews with users, who described their work procedures and their experience of using the networks.

The study showed that the visions and intended purposes of project networks do not comply with how such systems are perceived and used in practice. Users considered that project networks wasted precious time and were overly complicated. It was difficult to upload and structure documents and to describe them with correct metadata. Users also considered that it was difficult to find the information they needed and that it took time to log on, search for and open documents. They tended to use the networks as little as possible; if they did use them, it was primarily as a simple pool for storing documents that had already been approved. In other words, project networks were not used as active, dynamic communication networks but as passive, static archives. They did not support the intensive communication needed for problem-solving and decision-making processes. Instead, this vital communication was conducted through other channels, with information more likely to be distributed in real time rather than being stored and archived in the system.

These information and communication patterns are also highly prevalent in building production where such real-time distribution of information must function in mobile work environments which create other requirements for appropriate ICT support. No matter how much effort is put into the design and planning process, as soon as the production work on the construction site starts all kinds of problems and issues arise that call for immediate attention. In this constant reactive production environment, handling problem situations results in natural communication patterns that are dynamic, spontaneous and informal (Dainty *et al.*, 2006). The problems recognized with respect to information management and project communication on construction sites could possibly be explained by a partially misleading conception of what mobility is and what site-based mobile work involves.

For more than a decade ICT systems designed for stationary office use have been pushed out to the production environment, which has resulted in construction management teams being tied up inside site offices, at their desktop computers, for a large part of their working day. ICT implementation on construction sites has gradually forced production teams into partially unnatural and ineffective administrative work routines, due to the inflexibility and fixed nature of ICT systems. Extending these business systems to the construction site using wireless mobile computing devices will probably not be a sufficient solution to these problems in the long run. A legacy office-based system design will likely be forced on to a mobile

ICT platform that might need an alternative design to better fit the mobile work context. There are differences in how ICT is related to different work types. In office work, the computer is often the main tool for performing work, and functions virtually as the workplace itself. In mobile work the main job activities are regularly taking place external to the computer, often demanding a high level of visual attention and hands-on execution (Kristoffersen and Ljungberg, 1999). In mobile work environments such as construction sites, ICT-based systems play a supportive but important role, if they are designed according to the needs and requirements of the mobile workforce.

Modeling the totality

In the indicative studies described earlier it was found that communication was occurring on two levels at once. The formal, controlled exchange of documents took place on one level, while informal, interactive problem-solving took place on the other. Even though ICT plays a decisive role, communication cannot be viewed as a whole and is impossible to control it through formal tools. While ICT contains tools to enable us to keep track of the entire collection of information, it can also mean that a form of information anarchy prevails on the project. To explain this, we need to return to the basic question of how the 'as-yet un-built' can be visualized, communicated and understood among the participants involved in a project. Linn (1998) describes how technology based on 'pre-images' is actually a prerequisite for the construction of large, complicated buildings, forming architecture as knowledge:

> Images enable the pre-conception to be processed step by step. It serves as a work-piece in a visible process that is open to criticism. The various components can be kept apart and can be individually studied in a more analytical manner ... The situation is not unlike a game of chess: if the game is illustrated move by move, the consequences of individual measures and the choice of options become clearly visible and are available for action ... The significance of pre-image technology as a means of creation lies in the fact that it has enabled us to bring along a screen on which we may project and concretize the game and open it up move by move. The method has functioned extraordinarily well, has given rise to rich building traditions and has dominated the field for over four thousand years. It remains as useful today as ever, although we're now beginning to realize the potential of alternative methods more clearly than before.
>
> (Linn, 1998: 75, translated)

Computer modeling has added whole new dimensions to this knowledge technology:

A possible new knowledge technology may be glimpsed in the world of computer modeling. In the computer, an objectified virtual model can be created. It is not visible in itself ... The computer does not primarily create an image but models a 'virtual shape' which it is prepared to visualize in the form of an image displayed on the screen or on paper. This is where the computer has added a new step ... What is new is that the model's existence before the image has been split into two separate stages. After the model's first stage in the mental world the computer has inserted a virtual existence in which the model has been made collectively available. Several people can work with an identical model (at the start) and the changes they make can be referred back to the model. Its significance, therefore, is to a high degree communicative. So far, we have recognized only some of this new potential.

(Linn, 1998: 147, translated)

Immediate access

The vision of a common building information model (BIM) is very much alive, and great efforts are being made all over the world to realize this new means of sharing information; in fact, it has been of interest to researchers for the last 30 years or so. There is, however, still a long way to go before it sees full-scale use in architecture and building. The question of how practitioners can solve their communication problems in the meantime has in many cases simply been ignored. Much has remained as before, although with ICT as an additional factor to be managed in already complex situations.

The accepted practice for ICT-based project communication that has evolved over time is based on the use of web-supported project networks and the central storage of shared documents on project sites. This has given members of the project team immediate access to the information stored in the shared archive, but has reduced the flexibility and overall understanding of the project provided by the traditional approach to work and its practices. It is no longer possible to decide who is to receive what information at a given time. The information is available at all times, it is continually changing and project members do not wait to be given it. They obtain it from the easiest accessible source and hope that it is accurate and up-to-date. It is from this information that each participant creates his or her own 'pre-image' of the project. The difference between the old and the new approach to work is great. The project manager cannot control the images of the project that are being spread among members of the team.

Pressured by tight schedules and concerns over fees, everyone takes a chance on being able to complete their assigned duties at the last minute, which sometimes leads to near chaos. If, as in one of the above cases, after a year-long planning process a meeting has to be called on the day after distribution of the tender documentation in order to go through 600 corrections, anyone can see that much remains to be done before order can be brought to project communication channels. The point is that although ICT enables rapid communication and allows changes to be made at the last minute, it also creates new problems in such important areas as coordination, quality assurance and responsibility (Wikforss, 2006).

The ideal model of good project organization in the construction industry is the linear, hierarchical approach. The planning process is described in linear terms; it is divided into phases and is then successively broken down to an ever-finer level of detail. Everything seems to fit logically together. It appears that in construction contexts, the design and production planning processes are treated as a single process, even though the work involved in the design of a building differs significantly from technical planning and work on construction and detailed building solutions. The problem with this mechanistic way of thinking is that the ideas used to describe both the conceptual and actual construction of a building, from the finished whole down to the smallest detail, is also used to plan project organizations, human cooperation and the exchange of ideas between professionals – people who have very different educational backgrounds, knowledge and experience and who use different technical jargon.

Informal channels

ICT tools, too, are often put together in the form of systems which can be broken down into logical sub-systems and functions. When these systems are used in their intended context, the hierarchy of the organization, it turns out that they do not always produce the expected benefits but rather help to bring about the chaos witnessed by the participants in the project. The real exchange of information takes place via informal channels, where other forms of information and communication technology such as email, SMS messaging and mobile telephones are used, enabling direct contact between project members in network-like cooperation. The problem is that this communication behavior provides no possibility for ensuring the overall understanding and degree of coordination that a large project requires. How can a planned, mechanistic approach to systems be combined with a flexible, dialectical one so that it enables appropriate communication practices between interacting project members, just as a complex project demands? Dahlbom and Mathiassen (1993) discuss the importance of uniting these two perspectives:

> One of the challenges of systems developers is to understand and respect the Platonic nature of human knowledge and communication, and to understand the computer not only as a machine for processing data based on Aristotelian concepts but at the same time as a tool to support human beings in using and communicating Platonic concepts.
>
> (Dahlbom and Mathiassen, 1993: 37)

From dilemma to strategy

The question of how project managers should organize project communication involves much more than the choice of form and technology for representations of future buildings and whether it should be structured in two, three or four dimensions within a product and process model. A narrow search for standards for information deliveries, as the only solution to the serious communication problems encountered during the course of the project, obstructs many of the other factors that must also be handled by project managers. A variety of these factors can be identified in the indicative studies described above.

Formal and informal communication

The main question is how the project as a whole should be organized in order to facilitate both formal and informal communication (see, for example, Kraut *et al.*, 1990; Whittaker *et al.*, 1994). How can project managers achieve the flexibility of organization and method of work needed to enable project members to handle the many unexpected situations that almost by definition can be expected to occur during activities organized in the form of a project? How can a project organization and method of work be designed to support a combination of real time, interactive, ICT-supported problem solving and strict, quality-assured information deliveries? How can one facilitate rapid problem solving and direct contacts between the project members without disrupting the formal structure of the project?

Communication in the mobile work environment

The mobility of work is increasing in both the design and production phases of projects. Mobile work is often seen in relation to a place, for example an office or a desk, from which workers move away. Designing mobile ICT then becomes a matter of giving people the same possibilities in the field as they would have at their base. Mobility can also be a more fluid form of activity, where there is no such thing as a base. In fields like construction site work, mobility is an important component of the work itself. In these work environments people are mobile in relation to the kind of work to be performed, but they are not mobile in the sense of transporting themselves to some place to perform that work. This constant 'inbetween-ness' (Weilenmann, 2003) is an important part of genuine mobile work that results in contextual unpredictability and heterogeneity concerning job activities and their proactive and reactive assessments. This view on mobility poses new challenges for understanding what ICT is supposed to deliver in various job settings, as well as appropriate system design and use of the technology for different mobile work contexts.

Roles and incentives

The bases under which project members are taken on, their individual contracts and the distribution of their individual roles also affect communication. Attempts to define areas of responsibility too closely risk creating barriers between members of the team, who will reduce their individual contribution to communication within the project. Important information is lost and, in problem solving, participants tend to underperform when there is no incentive to provide information over and above the agreed deliveries. This also raises the question of what obstacles are created when new technical solutions for project communication upset the traditional distribution of roles. How can project managers deal with resistance to change, which is commonly encountered when different professional groups start defending their own interests?

Organization and management

As noted in the introduction, construction industry oriented information and communication research has until now concentrated on information modeling and standardization. In order to solve the practical problems that the industry is encountering, as described in the case studies, the perspective must be widened so as to include ICT from an organizational and management viewpoint (Sverlinger, 2000). How should one prepare, assess and decide on ICT strategies for differing purposes and financial conditions? How should one organize the merging of new enabling technologies and ongoing knowledge intensive activities? How should one organize ICT usage, and how should the overall operations be organized? Questions about the role of information technology in project management and its significance for knowledge formation, experience feedback and clear communications in project-oriented enterprises are becoming ever more central issues. It is also a question of how ICT affects the dynamic relationship between the individual and the project or company.

Usefulness and user acceptance

Achieving actual benefit from ICT tools is a matter of creating acceptance of the technology among the intended users through everyday usefulness in their ongoing work (Davis, 1989; Nielsen, 1993). The use of the ICT should not be conducted at the expense of other activities such as social collaborative processes, work practices or project management and leadership. One of the main challenges in this context is to understand the socio-technical gap of what is required socially within a work group and what can be done technically (Ackerman, 2000). It is important to understand how people really work in groups and organizations so that the introduction of new ICT systems do not deteriorate and distort the collaboration process and social interaction. If the technology does not serve and enhance these processes, it will be considered as an obstructive element for effective operations and project delivery, and will therefore not be used as planned. The technology has, therefore, to be designed as a supportive resource in everyday work that allows for intuitive and effortless use. In this sense, the usefulness aspect is about balancing the formal use, structure and functions that are embedded in ICT systems technology with the complex fluid and social nature of work practices and collaborative activities.

Implementation management

New changes, large or small, introduced in any project, corporation or industry will probably not turn into an immediate success. Tweaking both organization and technology will be necessary to achieve an appropriate configuration. The pieces of the puzzle do not fit together at the beginning and so it is through the continuous trial and error process of implementation (Fleck, 1994) that configuration of technology, communication processes and work practices can eventually fit the social and organizational context. This view on implementation as an enabling process for development involves continuous mutual adaptation between the technology and its environment, as well as recognizing the crucial role of the people inside the user organization. This collaborative adaptation process is necessary because technology rarely fits perfectly into the user environment (Leonard-Barton, 1988). Collaboration, communication and feedback between users and developers are often critical in achieving the proper fit between technology, organization and users (Rosenberg, 1982; von Hippel, 1988; Voss, 1988). User involvement in the technical development and implementation process therefore plays an important role in achieving long-term usefulness and benefit from ICT-based collaborative project communication tools.

Conclusions

The understanding gained from the case studies concerns the organization of ICT in project-oriented enterprises. The questions as such are of an interdisciplinary nature, since successful research in the field of project communication will derive from knowledge of developments in ICT along with profound understanding of the theories and practices of management and communication in relation to projects. One of the principle tasks will be to develop an understanding of the type of communication and information management that will be able to cross the many professional, disciplinary and geographical boundaries normally encountered in projects.

The improvement of project communication processes and technologies on different functional levels may change the organization of future projects and how business activities and work routines are designed, planned and performed. This can help, for example, in enabling just-in-time deliveries and the more industrialized and rational business processes that the construction industry is, in fact, striving to achieve. On-demand access and mobility of information, enhanced communication tools together with new ways of organizing and performing collaborative work could be important components of this development process. Full recognition of the need and determination to improve collaborative communication and information exchange throughout all project phases could have considerable impact on the industrialization process of construction projects. These issues have become a focal point for the construction industry. That is a welcomed change of attitude in a project-based industry that historically has seemed to have taken appropriate project communication practices for granted.

6 Transfer of experience in a construction firm

Mats Persson and Anne Landin

Introduction

The low level of transfer of experience achieved by participants in the construction process can be regarded as a weakness that reflects the lack of a natural forum for the distribution of information. Since each project undertaken by a firm is separated economically from every other, it can be difficult to link production and administration comprehensively. There is need of a system for reporting experience gained that is designed in such a way that all those engaged in the chain of tasks to be performed can access knowledge of the experience of others, both during a project and afterwards. If no such transfer of knowledge and experience takes place, there is the two-fold risk of the firm failing to take advantage of what has been learnt and of its making similar mistakes again.

Assembling information relevant to a project is part of the requirements of a management system generally, as expressed in the set of international standards ISO 9000/9001 (Quality management systems) and ISO 14001 (Environmental management systems), which are used in the construction industry at the level of the individual firm. The complexity of the construction process means, however, that special measures are called for if the collection of relevant information, including that concerned with experience gained, is to fulfill its purpose. The continual public debate regarding what takes place within the construction industry is considered by many to reflect flaws in quality assurance systems and the lack of well-functioning systems for collecting and distributing knowledge. There is good reason for the industry to identify ways in which the functioning of these two systems can be improved and these are examined in this chapter.

Theoretical framework

Value of knowledge

The stock market valuation of a firm differs from its book value. The difference between the two is often referred to as the firm's intellectual capital. The stock market valuation of one of the largest construction firms in Sweden is about two-and-a-half to three times as high as its listed equity. This is the result of the firm's intellectual capital being valued higher than its fixed assets. According to Sveiby (1997), a trend of this sort in stock quotations has been evident since the mid-1990s. Recent events may, however, change this condition in some way, although quite how it will manifest is hard to tell.

Intellectual capital can be defined as assets in the areas of knowledge, practical experience, organizational technology, client/customer relations and professional skills, which serve to provide a firm with competitive advantage in the marketplace (Edvinsson and Malone, 1997). Intellectual capital can be regarded as knowledge that can lead to profit (Sullivan, 2001). Human capital is the value of that knowledge held by people, which is to a large extent tacit knowledge. According to Sveiby and Risling (1986), knowledge management represents the creation of these non-material assets.

Sveiby (1997) regards the human being as at the center of the knowledge-based firm, rather than its production processes or financial capital. A knowledge-based firm is thus quite different from an industrial firm, just as it also differs from a construction firm. Much of what characterizes an industrial firm is also typical of a construction firm, although there are certain aspects of the latter that are also typical of the knowledge-based firm. This can be seen in terms of their both handling information and regarding knowledge as important. Dancy (1985) emphasizes the difficulty in defining knowledge and traces developments that the concept has faced over the years. Holden and von Kortzfleisch (2004) and Dancy provide examples of both the rational and the empirical direction in conceiving of knowledge. Rationalists declare that knowledge can be derived through the use of mental constructs in the form of concepts, laws and theories, whereas empiricists claim that knowledge can only be generated on the basis of experience and observations.

Concepts of knowledge management

Polanyi (1967) basically places an equality sign between knowledge and the ability to know, such that 'knowledge is an activity which would be better described as a process of knowing'. Both Polanyi and Drucker (2003) take up the idea of knowledge always being found in a social and political context. Since knowledge is bound to the individual it is basically silent (tacit). Tacit knowledge, including its transfer, content and the processes involved, is poorly understood (Foos *et al.*, 2006). There are numerous examples of attempts that have been made to gain an understanding of the knowledge of the expert by documenting, as thoroughly as possible, all of the expert's reflections and thoughts. Those who have tried to obtain an adequate grasp of knowledge in this way have likewise witnessed the difficulties involved. It is extremely challenging to describe one's tacit knowledge in order to make it explicit.

According to Nonaka and Takeuchi (1995), Plato defined knowledge as a 'justified true belief'. They describe tacit knowledge as 'deeply rooted in actions and in the individual's engagement in a particular context'. Organizations differ in the tacit knowledge they possess, it being a function of the experience an organization has amassed over the years. It is impossible to communicate tacit knowledge simply by use of diagrams and printed material. Marcotte and Niosi (2000) maintain that tacit knowledge, according to Polanyi's definition of it, is most frequently found in scientific and technical areas. Polanyi (1967) declares that no completely explicit and codified knowledge exists at all, since behind every bit of explicit knowledge there is also tacit knowledge about which one is unaware.

Davenport and Prusak (1998) describe a dynamic process of attaining knowledge involving four stages: generation, codification, transfer and realization. Generation consists of all processes concerned with the acquisition of knowledge. Codification involves the processing of knowledge in a manner that puts it into a structured format, i.e. one that allows it to assume the role of explicit knowledge. The transfer of knowledge involves its communication to other people or conversion to other forms. In the final stage, that of realization, both the recipient of knowledge and the organization to which the person belongs profit from it, since the information has value for both.

The transfer of knowledge within and between that of the tacit and the explicit types can be divided, according to the SECI model (Nonaka and Takeuchi, 1995) into knowledge transfer of four different types: socialization, externalization, combining and internalization. There are different starting conditions that lead to knowledge transfer. The first of these is *socialization* which, in order to start, requires that a space for interaction be created – one that enables participants to communicate their experiences and ways of looking at things. The second is *externalization*, which is supported by a meaningful 'dialog or collective reflection', that helps participants to express their tacit knowledge which is otherwise difficult to communicate. The third is *combining*, which is activated by network activity concerned with such knowledge as that of new products, jobs or leadership systems and involves people from different parts of the organization. Finally, *internalization* is brought about by explicit knowledge being utilized within the framework of learning by experience. The chain of knowledge transfer thus described can be regarded as a spiral that rotates turn after turn within the organization.

Nonaka and Takeuchi (1995) point out that individuals generate new knowledge and that an organization needs to learn to mobilize knowledge accumulating at the level of the individual. The tacit knowledge that the organization has at its disposal can be increased by means of the four basic methods of knowledge transfer described in the SECI model. The spiral type process, mentioned above, can lead to both tacit and explicit knowledge developing and expanding. This learning spiral of the organization begins at the level of the individual through interaction by individuals (i.e. socialization). The level of knowledge then continues to rise through the processes of externalization, combining and internalization.

Other aspects of interest

Examples have been presented of different ways in which the acquisition of knowledge can be studied. Nonaka and Nishiguchi (2001) describe the development of knowledge in terms of an ontological dimension of *individual–group–organization–sector*. One can ask where on this scale knowledge of a particular sort lies and where bits and pieces of knowledge of various types are located. How can knowledge be moved between different points or regions on the scale? One problem is that the scale is not linear. Often an individual is a member not solely of one group, but also of an organization and a sector. Knowledge of a particular sector represents knowledge that those within the sector generally possess and about which they are in agreement (Persson, 2006).

Another way of dividing knowledge up is in terms of the shifting perspectives a temporary organization can have, which one can liken to the different phases of the construction process, namely those of designing (or creating), producing (or carrying out) and administering (or storing) and the knowledge that relates or can be applied to them. Carrying out a construction project is often described as a kind of relay race in which information is passed from one actor to another (Söderberg, 1994). Such information or knowledge can be of a temporary nature for many of those participating in a project. They may only need to apply their knowledge during that part of the distance to be covered in which they are holding the baton. The anchor runner hopes to be in a good position when handed the baton. Similarly, some of the information available in a temporary organization at a particular time may not be needed until later, when it can prove valuable. Knowing how knowledge from the project being carried out will be dealt with later, or will affect those for whom the project was undertaken, can be important. The needs, wishes and expectations of those who are to later own, use and maintain the object being created are conveyed in many separate steps and stages by those who are working on the project through its various life cycle phases. An alternative way of looking at this is to consider the various interest groups involved in a project, such as those of the owner, end-user, architect, other designers, site manager, supervisor, foreman, tradesman and administrator. This chain of actors often has side-chains and recursive loops that can make a construction project a very complicated undertaking (Persson, 2006).

A third way of dividing up knowledge is from the perspective of a permanent organization, which in this case concerns the question of how the firm manages the knowledge it has acquired from the training and experience of its employees and the projects it has carried out. There is also the matter of how this knowledge can be used in a manner enabling the firm to be as effective and as competitive as possible, for which reason the firm endeavors to develop its management system, the competence of its people and the processes involved in its work.

The environment in which knowledge is embedded, such as expectations placed on a particular sector of the economy or pressures for change, can be added to these three perspectives. Over the past decade, new ways of organizing construction projects, based on the idea of greater cooperation and openness, have come to the fore. Considerable emphasis has also been placed on making the construction process as rational and effective as possible (Persson, 1999).

Permanent and temporary organizations can be seen as examples of the dimensions of matrix organizations. Parallels can be drawn with various discussions of matrix organizations in the project management literature, such as that found in PMBOK® (PMI, 2004). Assembling knowledge and sharing experience are particularly important in the permanent or hierarchical organization. There, the possibilities are particularly good for the extensive sharing of experience with colleagues and with others (Persson, 2006).

In temporary organizations, knowledge of a specific project and the use of routine checklists often play a central role. Knowledge and information often need to be handed on to the next actor in a kind of relay race as mentioned earlier. Some stakeholder groups, such as craftsmen on a construction site, need to assimilate information of momentary importance, such as written material to be internalized and what they are told on the site by those in charge.

It can be difficult at times for craftsman to understand from explicit sources how a particular step in the construction process is to be carried out. Tacit knowledge can play a major role under such circumstances, by doing what one is accustomed to do without studying drawings or written materials first. The readiness to work in this way (i.e. figure out things on the spot) can be a positive trait, especially when no drawings or descriptions of the exact procedures to be carried out are available. Yet, it can lead to insufficient precision and result in quality requirements not being met.

Research project

The research covered in this chapter is based on a case study, primarily the analysis of documentation in a construction SME. The rationale for this choice was that a large organization might more easily provide ambiguous information. The management of knowledge by SMEs was seen to differ from that in larger organizations, not least because of the limited resources that SMEs might have available (Desouza and Awazu, 2006).

The overall aim of the study was to analyze how knowledge and experience could best be built up and made continually available to those in need of it, with a system designed to serve these ends as effectively as possible. A specific objective of the study was to investigate how a particular construction firm deals with the knowledge potentially available to it and how the organizational work of the firm supports its knowledge management efforts (Persson *et al.*, 2006).

Methodological considerations

The study makes use of case study methodology and document analysis. Data had been collected over a six-year period from a close involvement with the firm. Concurrent with the study was the formation of a reference group consisting of people with long experience of construction firms and a deep understanding of the circumstances affecting them. The members of this group were selected from different organizations. An attempt was made in the study to determine the types of knowledge and information that appeared to be the most important to the firm and the impact these had on the organization's functioning, an approach adopted from Kolb (1984).

The management system used in the company – based on ISO 9001:2000 – served partly as the basis for deciding on the measurements and assessments to be made within the organization. ISO 9001 contains requirements and principles concerning the organizational work that needs to be done to assure the quality of products or services and the improvement and learning that take place within the organization. Greater openness to contact within the organization is also a goal (Deming, 1986) with programs to be organized for breaking down barriers and increasing people's engagement. Individual learning (Senge, 1990) is emphasized. One of the eight fundamental principles for quality management in ISO 9001 is that of continual improvement. This is defined as 'recurring activity to increase the ability to fulfill requirements'. According to ISO 9000:2005,

> the process of establishing objectives and finding opportunities for improvement is a continual process through the use of audit findings and audit conclusions, analysis of data, management reviews or other means and generally leads to corrective action or preventive action.

The document management system (DMS) of the firm was implemented as far back as 2000 and is an important source of information. All working documents are held there, including modules for the following.

- Document templates and project documents' storage.
- Handling and storage of drawings.
- Time schedules.
- Budgets and cost estimates.
- Time spent working on various tasks.

- Workload and orders.
- Follow-up.
- 'Non-conformities' and change orders.
- Diary of work undertaken.
- Library of reference literature on the construction industry.
- Other documents and minutes of meetings.

One heading used in the DMS system is that of 'non-conformities and change orders': it involves database information stored locally at production sites and at the head office. Under the heading of library, documents and minutes covering management reviews and 'experience meetings' were collected and made available for examination.

Twenty-three construction projects in which the firm had been engaged were selected for investigation. The total number of non-conformities found was 962, covering 12 of the projects, there being no non-conformities noted for the remaining 11 projects. Records of all management reviews conducted by the firm were stored in the DMS. Thirty-four of them were completed in accordance with the certified quality management system and were examined more closely.

Research results

Overall analysis of reported non-conformities

The firm employs a very broad definition of the term non-conformity, which includes adjusting the scope of the contract from what was planned originally. Many of the recorded non-conformities give the impression of being what is termed, within project management (PMI, 2004), 'changes in scope' rather than failure of the product to meet the requirements placed upon it.

The DMS lacks clear and distinct terms for certain matters and this means that a given term may be lacking in specific content and/or intent. The concept of non-conformity should probably be reserved for what ISO 9000 designates as *non-fulfillment of a requirement*, with work stemming from a change in product scope referred to in other terms, for example change order – something that is readily understood within the construction industry. The term non-conformity can also be associated with critical comments made at the time of an inspection, yet the DMS system provides no information about the results of such inspections. This means that a total picture of the non-conformities is lacking, which reduces the chances of assembling knowledge and sharing experience.

Registering non-conformity is an example of externalization in which tacit knowledge becomes explicit. The use of appropriate terms makes it easier both for explicit knowledge to be collected and for knowledge to be internalized. The lack of an adequate system for proper codification resulted in externalized information being unusable for further combination or for subsequent internalization.

In discussions within the reference group, it was suggested that the time factor tends to place emphasis on being as productive as possible. Construction site managers were seen as under particularly strong pressure and face a difficult work situation. Financial results and client/customer satisfaction were both given high priority. Much less emphasis was placed on the accuracy and adequacy of the classifications and codifications.

Another possible reason for non-conformities sometimes not being documented by the firm and for the limited attention directed at them could be to do with the work culture, which is not bureaucratic in terms of regarding documentation as all-important. There is, instead, a culture in which conversation, discussion and social networks are regarded as the principal ways of spreading knowledge and information (Gluch, 2005), although this aspect was not investigated systematically.

Management reviews

Analysis of records – typically minutes from the management reviews – showed that process-oriented questions tended to dominate (77 percent). The minutes revealed the presence of certain basic tendencies in terms of the conduct of the reviews. Questions placed on the agenda tended to remain there until a final answer was found. Agendas also appeared to stimulate discussion of the matters with which they were concerned. Questions raised in internal audits of the quality management system were also taken up in management reviews. Half of the questions dealt with in reviews concerned routines, checklists and approaches to consultations. These are important topics in the quality manuals, both in connection with the tacit knowledge presented within the firm and for the management system generally. The overall impression of management reviews was that tacit knowledge possessed by attendees tended to be externalized there and then to become explicit knowledge that found expression in various documents, in the approaches taken to consultations and embodied in the management system.

Some of the agenda items in the reviews involved reporting of information that was not analyzed further. This applied above all to non-conformity reports in which the facts as presented were not regarded as calling for any particular measures to be taken. This was because the non-conformities in question represented changes either already approved under the contract or as extensions to it.

The wording of minutes was precise and accurate, thus meeting the requirements of the certifying organization and ISO 9000, all of which is very positive. The minutes also served as useful checklists to ensure that nothing of importance would be omitted. At the same time, the form and organization of the minutes makes it difficult to combine different aspects of the information contained in them in a way that enables it to be incorporated into the DMS system. Both the form and the somewhat limited availability of the minutes make it difficult for employees to gain ready access to their contents and so are not able to apply the lessons learned. Yet, the firm makes an effort to present information to employees concerning the conclusions of the reviews in a way that is aimed at helping them internalize the issues involved.

Meetings for the sharing of experience

Those attending meetings intended for sharing experiences were primarily clients and those in charge of the project in question, consultants from outside and a few suppliers. The firm's senior managers reported that the interest which clients had in coming along appeared to be only lukewarm. After each meeting, minutes were written up to demonstrate to potential clients that the firm worked with matters of quality in a systematic way as well as contributing to the overall marketing of projects. Even so, what the firm considered most important was identifying anything with which clients or others were dissatisfied.

It can be argued that such a firm would be better holding meetings of two types for the sharing of experience on individual projects – one dealing with experience reported primarily by external sources and the other for experience reported by internal sources. The meetings currently held are of the external type, concerning how individuals and firms collaborate and utilize experience, and the way in which contracts are drawn up. In contrast, meetings of the internal type, where those participating were all from within, discussed matters of profitability, budgeting and other financial matters, as well as whatever weaknesses there appeared to be in the firm's organization. Clearly, these are matters that are better discussed in the absence of external people.

Minutes of meetings covering the sharing of experience are examples of the externalizing of experience. Just as in the case of management reviews, information by itself may appear of superficial value and may not be particularly accessible in depth; but it does satisfy the need to provide something to those who need to report on the functioning of the quality system.

Discussion and conclusions

Support for knowledge acquisition through use of quality management

The case study firm tended to use the term 'non-conformity' to cover change orders. Parts of the system for registering non-conformity could, in principle, be used for adding information from external sources, although no evidence of this use was found. In the minutes of the reviews and of meetings for the sharing of experience, non-conformity tended to be handled in a rather undifferentiated way. Analysis of non-conformities concerned mostly how issues might be taken up in discussion with the client. In management reviews, questions of whether non-conformity reflected poor planning, inadequate coordination of the work or failure to meet standards and the like were sometimes discussed (i.e. non-conformity relating to errors), but not always.

Non-conformity due to changes in design or other sources of extra work could have to do with special situations that developed, the wishes of clients or issues arising with suppliers. For a construction firm, approved changes and extra work need not be a problem. The firm does, however, need to document and control them and ensure that it is paid correctly. It is possible that a more thorough analysis of the procedures for managing changes would show some weaknesses in the processes and routines involved. Information should also be present within the system so when properly integrated could be communicated through socialization or externalization, enabling extra costs arising from changes to be reduced.

Non-conformity as defined in terms of ISO 9000 should represent an ideal point of departure for learning and for the transfer of knowledge. A permanent firm has the opportunity to increase its knowledge continuously with the aim of avoiding mistakes. Non-conformity can be noted in management reviews, which primarily concern matters of documentation, routines and procedures of a general nature. One reason why it may be difficult in a business organization to deal adequately with both major and minor problems is that there is much greater profit (and hence motivation) for a firm to use limited resources on business matters of major import than concentrate to an inordinate extent on overcoming weaknesses in areas of little financial import.

Management reviews were concerned in part with questions of the quality management system. Questions taken up for discussion resulted in updated routines, checklists and other documentation. Questions of training and of the need for special courses were also discussed and followed up. This made such meetings a good forum for considering matters relating to the assembly of knowledge and the sharing of experience generally. It appears uncertain, however, if the firm has a clear-cut strategy or any concrete plans for assembling the knowledge it has acquired and then disseminating the results of the experience gained from that throughout the organization. Neither the minutes of these meetings nor of those for the sharing of experience contain any appreciable information regarding the development of production methods and products. This is definitely a weakness in terms of knowledge acquisition and experience sharing. It should be of interest to the firm to follow-up developments in both these areas with the aim of enhancing the firm's effectiveness and competitiveness. The same comment can be made concerning the firm's purchasing, cost planning and logistics functions. These important processes appear, according to the minutes of the management reviews, to be insufficiently considered.

Reliability and validity of the results

The study illustrates, with the help of examples, how the acquisition and assembly of knowledge and the sharing of experience can function within a firm. The firm in question was selected because it has a quality management system in accordance with ISO 9000. It is regarded very positively and has received an award for its work in this area. Accordingly, the firm can be expected to have maintained a high level of quality output compared with other firms within its sector. The study, nonetheless, reveals some particular weaknesses in the firm's quality management system. It is reasonable to expect similar weaknesses to be found in other firms.

Acknowledgments

The authors wish to thank SBUF and FoU-syd for their financial support, which made this study possible. We also like to thank all the staff of Thage Andersson Byggnads AB for their time and their generosity in sharing thoughts, experiences and documents. Finally, we thank all others who, at different stages in our study, gave their time for interviews and discussions.

7 Autonomy and innovation in construction teams

Niels Haldor Bertelsen

Introduction

Poor quality, many errors, low efficiency, high prices and low credibility characterize the Danish construction industry's image in the eyes of the general public, authorities and government. At the same time, the industry is going through tremendous change. Production on construction sites is shifting toward industrial methods. New forms of partnership and models of organization are being introduced in and between enterprises. A multitude of new materials and products are replacing old familiar ones, and the industry is under pressure to use digital models and 3D-visualization in communicating with end-users and collaboration partners. In addition, Danish authorities have a vision of more user-driven innovation in all parts of society.

How do specialist contractors and small firms respond to these challenges? What is the role of the skilled worker in this development? How can firms and workers contribute to a solution to these problems and improve their position and opportunities for development? This chapter attempts to answer these and related questions. As the objects of our research, we have chosen the small construction firm and skilled workers on the construction site. This study is a first step, aimed at finding a structure and model for national efforts to increase competence in innovation. The study is also intended to encourage the testing of key measures, competences, better collaboration among trades, education and research in the specialist area of bricklaying.

Literature review on innovation models

Literature reviews have been conducted on several occasions as the need arose in the course of a series of small case studies which this action research has been undertaking. The cyclical approach between literature review, case study and reflection over key themes is an approach that Barrett and Sexton (2006) have utilized. They consider that the 'hybrid approach loosens the disadvantage of grounded theory, that it can be unduly limited by the cases and the researchers'.

Segments, categories and drivers of innovation

Literature and textbooks on innovation have different foci depending on whether the author is an economist, technical specialist, designer or sociologist. There is also a significant difference in the language used by researchers, management consultants, developers, enterprise leaders and ordinary employees. In a comparative study of cultural differences between the Danish and the British construction industries, Hancock (2000) shows that there is a greater difference of culture at the level of trades and specialists than between the two nations. Understanding these differences is crucial when attempting to transfer experience from one trade or sector to another.

Ingvaldsen *et al.* (2004) compared construction productivity in the construction industries of the Nordic countries in an attempt to suggest where improvements might be made. This comparative study adopted two approaches: productivity and innovation from the perspective of the contractor and from the value-oriented perspective of the owner/client. The study also discussed differences between new building and renovation, and between house-building and other types of construction work. The study proposed that future development should differentiate between segments of the industry/sector and the actors involved. In the course of the study, a gap between national statistics and day-to-day construction practice as experienced by the parties involved in projects was identified.

The division between the different segments formed a topic for further discussion (Bertelsen, 2007). It was concluded that 'process and product' should be divided into owner and end-user, planning and design, construction site and industrial production. This division was seen to allow four competence categories to be covered: lean construction, information and ICT, quality and defects, and economics and life cycle cost (LCC). It was further suggested that effort should be expended in improving national statistics and creating common methods for case studies, since the former were erroneous and no common practice existed for the latter.

In a more extensive literature review of innovation in small, project-based construction firms, Barrett and Sexton (2006) emphasize the importance of the role played by these firms in industrial markets and in technological change. They conclude that owners of small firms have a dominant role in driving innovation activities and that there are two modes of innovation: mode 1 – single project and cost-oriented client relationship and mode 2 – multi-project and value-oriented client relationship. Here, the division between project/firm and cost/value, as implied in Figure 7.1, seems better able to describe the differences between modes.

Barrett and Sexton (2006) also emphasize the wish for a more holistic and dynamic approach combining the following drivers: technology, market and owners' resources. A knowledge-driven approach could also be an option in this connection, if one believes that research, development

	Cost and process	Value and product
Single-project	Mode 1	–
Multi-project	–	Mode 2

Figure 7.1 Different innovation modes (after Barrett and Sexton, 2006).

and education can be drivers of innovation. In their conclusion, Barrett and Sexton stress that policies for large construction firms are not necessarily appropriate for small construction firms, and vice versa, and that the research focus on innovation in small, project-based firms is very much in its embryonic stage.

Case studies on the renewal of old, multi-storey housing in Denmark have introduced the hypothesis as to who the drivers of change might be (Bertelsen, 2004). First, it was stated that a certain difference must exist between the poorest and the best firms in the market. As Figure 7.2 illustrates, frontrunners create a pull in the market through innovative thinking and continuous improvement. There is constant pressure from the market on the poorest performing players, forcing them to respond either with an innovative 'jump' (jumpers) or by leaving the market (losers). Our hypothesis is that motivation can be used to woo frontrunners to pull the market forward, and that regulated pressure on potential 'losers' and 'jumpers' will ensure the right difference of supply.

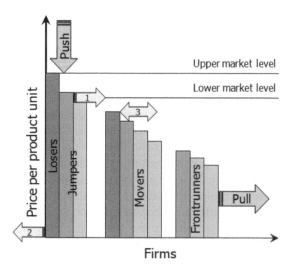

Figure 7.2 Different types of player (Bertelsen, 2004).

The above is summarized in Table 7.1 below showing a list of the most important segments, categories and drivers. These are expected to be useful in one way or another in further work on building a model for understanding autonomy and innovation in construction teams.

Models and tools for building an innovative environment

Barrett and Sexton (2006) have proposed a generic innovation model in five elements (Figure 7.3):

1 innovation focus and outcomes;
2 enhanced performance;
3 context in innovation;
4 organizational capabilities for innovation; and
5 innovation process.

They concluded from their study that there are eight research gaps concerning innovation in small, project-based construction firms; for example, gap 8, 'Is the process of innovation rational and/or behavioral in nature?'. They propose additional factors for use in the innovation model: organization of work, people, technology and business strategy/market positioning (Figure 7.4), which cover both mode 1 and mode 2 in Figure 7.1. The factors resemble the four drivers in Table 7.1.

Personal mastery

Senge *et al.* (1994) define personal mastery as exceptional skills or 'master of a craft', and one of the fifth disciplines in developing a learning organization.

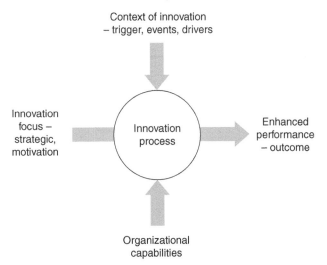

Figure 7.3 Generic innovation model (after Barrett and Sexton, 2006).

Table 7.1 Different segments, categories and drivers of innovation

National actors	Building	Process and product	Competences	Firms and project	Innovation driver	Types of actors
Public and national	Housing, new	Owner and end-user	Lean construction	Big firms, multi-project	Technology (external)	Losers
Building sector	Housing, renewal	Planning and design	Information and ICT	Small firms, multi-project	Market (external)	Jumpers
Other sectors	Others, new	Construction site	Quality and defects	Single projects	Resources (internal)	Movers
Research education	Others, renewal	Industrial production	Economics and LCC	Individuals and teams	Knowledge (internal)	Frontrunners

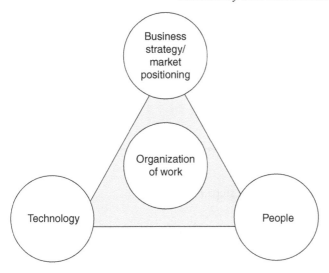

Figure 7.4 Organizational factors in innovation (after Barrett and Sexton, 2006).

Having a clear picture of one's current reality and a vision will produce a 'creative tension' in the innovation process that looks for ways to move closer to what we want. The tension between reality and vision can be likened to a rubber band and forms a central element in personal mastery. Current reality corresponds to 'innovation focus', vision corresponds to 'enhanced performance and outcome', and the 'rubber band' is the external tension that creates the 'innovation process' (see Figure 7.3).

Senge *et al.* (1994: 226–9) stress that personal mastery also includes being able to handle 'both a reactive response to events and a creative response to reach the future and vision you want'. This statement accords well with Barrett and Sexton's (2006) ideas about the rational and/or behavioral nature of innovation. The rubber band can be envisioned as extending also to the 'context of innovation' and 'organizational capabilities', providing a double tension for the 'innovation process'. At the general level, Senge *et al.* (1994) complement Barrett and Sexton's (2006) specific study of small construction firms. The former offer several proposals for different tools to be used in the innovation process to help close the eight gaps in the model identified by the latter.

Storytelling is one of these tools, and provides opportunity for describing the correlation between cause and effect in an entertaining manner as, for example, a springboard for learning and innovation. Senge *et al.* (1994) suggest expanding the tool to four levels: events/context of innovation, pattern of behavior, systemic structure and mental models. Through pattern of behavior, the process, trends and responsibilities can be described. Furthermore, connections and relations may be determined in a systemic structure, as a part of systems thinking – the second element of the fifth discipline. In the mental model, the third element of the fifth discipline, the ladder of interference is described in the following steps:

1   observe data and experiences;
2   select data;
3   add meanings;
4   make assumptions;
5   draw conclusions;
6   adopt beliefs; and
7   take actions.

Reflection is a loop back from steps 6 to 2.

Research objectives and methodology

Objectives

The target group for the research is small construction firms and skilled workers. The research objective is to develop a 'grassroots' model for innovation and to test it on small bricklaying firms and bricklaying gangs. The test serves as an illustrative example for other trades, small construction firms and skilled workers. The intention is to build a simple model for innovation that can be applied by skilled workers and supported by a local innovation network centered on the technical schools. A further objective of the research is to find answers to four key questions, which form the subject of three sub-studies. The findings from the first part of the sub-studies are reported here.

Methodology

The approach can be characterized as inductive research, which has been performed in close collaboration with construction firms, skilled workers and technical schools. A number of small case studies of construction teams and small construction firms were conducted, iterating between development of the model and literature reviews. This iterative approach has led incrementally to a number of insights that have been tested on different actors in the sector (construction firms, skilled workers, architects, engineers, researchers and teachers) through lectures and discussions in working groups. In this way, simplicity and clarity have been sought for the model to make it easily applicable and effective for the target group, while at the same time basing it on a research foundation. This inductive method dates back to 1637, when René Descartes described it in *Descous de la Méthode* [The Method].

Small construction firms and skilled workers were chosen as the target group from the beginning, because they play an important role in the construction industry. At the same time, we could see that research was sparse in this area. It was then decided to look into the opportunities for developing autonomy in teams and cross-disciplinary collaboration on the construction site, as an alternative development path for specialization. This also provides a chance to assess future competence development for the target group. Finally, it was decided to focus on bricklayers as the trade example. This choice was made in collaboration with the labor union, because the trade is known for its traditions as well as the need to review its practices.

The literature reviews and our understanding of the needs of the trades led us to the following questions:

1 Can a simple 'double rubber band model' be developed (based on Figure 7.3) and experience brought in from other industries to enable small construction firms and skilled workers to improve their business and competence?

2 Can frontrunners be developed (see Figure 7.2) so they may serve as a locomotive for development and learning on the construction site, in the firms and in schools?

3 Can the technical schools become local development centers and an innovation link between research and small construction firms and skilled workers?

4 Can knowledge be the fourth innovation driver (see Table 7.1)?

5 Can small construction firms and skilled workers achieve competences in all four modes (as in Figure 7.1) and in this way contribute to speeding up industrial development and create better productivity and value for clients.

These themes will be analyzed through the three sub-studies, of which the first two have been completed and the third has been launched.

The first sub-study is an analysis of autonomy in construction teams and future competences for skilled workers. This study was carried out in three parts. The first part consisted of ten workshops with a group of experts (researchers, architects, engineers and teachers) and union/organizational representatives to set the framework for future development. The second part was a study of general theories and experience from teamwork, communication and autonomy aimed at skilled construction workers, which was to show experience from other sectors applied to the target group. The third part consisted of on-site interviews with four highly qualified skilled teams and master craftsmen from different regions of Denmark. Based on findings from these three inputs, the idea was to provide an overview of required future competences for skilled workers and to point toward key areas for development.

The second sub-study was of small bricklaying firms and strategies for developing efficient methods for controlling processes, logistics and quality in teams and in small construction firms to illustrate the first step in an innovation process. The third sub-study will start a long-term test and development project for the bricklaying trade. Development will take place through local competence centers in collaboration with research institutes, design schools, business schools and local, innovative firms of bricklayers and bricklaying gangs.

Research results and industrial impact

The results of the first and second sub-studies have been reported in Bertelsen (2005) and Gottlieb and Bertelsen (2006). The third sub-study has been initiated but results have not yet been reported, so only preliminary experience has been included from this illustrative test for establishing an innovation network for bricklaying firms and skilled workers.

First sub-study – autonomy in construction teams

The first sub-study was launched through the following three activities:

1   ten workshops with a group of Danish experts and organizations;
2   literature review – references are not included in this chapter, but can be seen in Bertelsen (2005); and
3   interviews on autonomy for four construction teams.

It was clear from the ten workshops with the experts and representatives from the labor union and other organizations that their knowledge was primarily on the framework conditions for construction teams and skilled workers. They knew something about education, collective agreements, project material, client requirements and legislation (points 6, 5 and 4 in Figure 7.5), but they knew little about the specific work of the teams and their development opportunities (points 1, 2 and 3 in Figure 7.5).

The main conclusion from the workshops was that participants had indicated ten important development areas, five of which were aimed at technical skills, one at cross-disciplinary collaboration, two at self-governing teams and two at external frames (see Figure 7.5). They advised against starting development concerning public norms and agreements because this would take a long time and it is likely that there would be very little impact.

The literature review showed that there is a lot of general experience of autonomy, collaboration and innovation, but little of this had been adapted to the special needs and conditions of small construction firms and teams. This matches well with the experience of Barrett and Sexton (2006).

The four teams and firms were selected because they were amongst the best, as well as representing different trades and regions in Denmark. The interviews showed that skilled workers had very good technical competences,

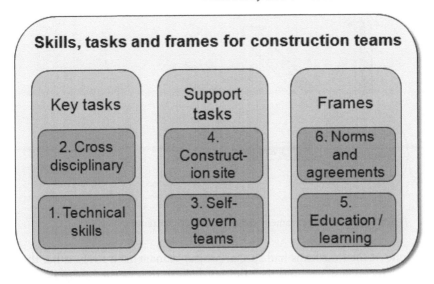

Figure 7.5 Important skills, tasks and frames for construction teams and workers.

they worked autonomously and so the firms wanted to give them more responsibility. One team was particularly interesting as it had developed new tools which encouraged more efficient processes and improved health and safety. This development was supported by the firm and many of the tools have been developed further and are now in use by other firms.

Overall, this sub-study showed that developments in the future will very much build on the high technical competences possessed by skilled workers, and there is a trend toward more autonomous teams, although developments related to cross-disciplinary collaboration are not moving so quickly. Grassroots innovation in small construction firms is possible, fueled by innovative skilled workers and master craftsmen, and this is being practiced in many places in Denmark. There is no doubt, however, that there is a lack of education and knowledge on how this can be implemented and improved (see Figure 7.6).

Second sub-study – innovation of process control in small firms

Four small bricklaying firms took part in this sub-study; they were selected as the best in a local area north of Copenhagen and they represented different sizes and organization. A simple illustrative method of process and quality control for small project-based construction firms was described to them and proposed as a future management tool. This first drawing board prototype of the tool was used at the interviews as an alternative to current practice, and on the basis of this the firm's development opportunities and wishes were discussed.

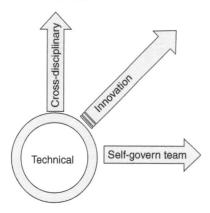

Figure 7.6 Four development strategies for competences.

None of the four firms had special project management tools, other than their financial systems, and they did not believe they needed them because they were not required by their customers; besides, the firms felt they could manage via mobile phones. All firms wanted more education and training if they were to use the new tool anyway, and the tool should be fully developed and provide rapid results.

The four firms were thinking primarily in mode 3 (see Figure 7.7) and although they wanted results from the project management tool, it was hard for them to discuss the business aspects – see mode 1 in Figure 7.7. This was despite financial and quality assessments of the firms' work showing that they did have a need and that there were financial benefits to be made. Consequently, it was concluded that future development of the tool should follow the specific order of events in Figure 7.8.

The response to pressure from the market was reactive and non-creative, and three firms (firms A, B and C) had sought out a niche where demands were limited, nicely befitting their low competitiveness. It was surprising that among the four there was no frontrunner or even one that had a goal of moving the firm to a higher level of competitiveness. This leads to the conclusion that a market push and a technology pull (see Table 7.1) cannot be considered adequate drivers of innovation, and that the resources of

	Business	Technique
Tool	Mode 1	Mode 3
Communication	Mode 2	Mode 4

Figure 7.7 Proposed future development steps 1, 2, 3 and 4 (focus is on mode 3).

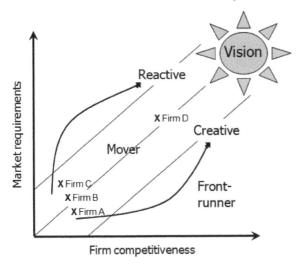

Figure 7.8 The four firms in the rubber band 2nd tension.

owners and profit are protected rather than applied for innovation. Furthermore, it is felt that an innovation pull from role models in the trade or sector (frontrunners) is required, and that there is a need for knowledge about and training in innovation for all modes as shown in Figures 7.1 and 7.7.

Third sub-study – establishing an innovation network for bricklayers

In 2006, a project was launched by employers, employees and technical schools in cooperation with researchers and with total support of €300,000 from various sources. The first two innovation networks have been established and they have implemented education and training in innovation and development of new internal walls and bathrooms. Preliminary experience has been very positive.

- Frontrunners among skilled workers, small construction firms and teachers constitute efficient drivers of innovation.
- The local innovation network is an efficient link between research and small construction firms, but there is a lack of competence in innovation and teaching in addition to technical skills, both in the firm and on site at the master craftsman level.
- A simple innovation model based on 'double rubber band tension' as a framework for the innovation process, which is easy for the target group to use, seems to be a practicable solution; however, experience of use in practice by the target group is lacking (Figure 7.8).

- Many development elements have been put into play concurrently in the project providing many synergistic effects, which are necessary for usability. It is also important to have broad skills and a pedagogical approach, characterized by adaptability, rather than specialization and heavy research.

Conclusions

The bricklaying trade and trade schools have welcomed the new innovation network which is headed by local technical schools and intended to support the long-term development of the innovation competences of small bricklaying firms. It is anticipated that testing this grassroots model for user-driven innovation on a selected trade will spread to small construction firms in other trades, e.g. carpenters, floor layers and plumbers. The new innovation network is a bridge between research, technical schools and small construction firms in an equal, collaborative relationship, supporting the innovative frontrunners and the dissemination process.

This chapter has sought to explain four key themes in a model for innovation in small construction firms. The concept is a simple 'double rubber band model', providing creative tension to the innovation process; it seems also to provide the outer framework for development of the area. Frontrunners appear to be an essential driving force at the construction site, in firms and in technical schools. Furthermore, it appears that technical schools have a need to develop their innovation competence and to aim for balanced development collaboration with firms and construction sites. Following the first tests, it looks as if the target group of small construction firms and skilled workers will be able to carry out accelerated development in different modes toward the goal of industrial production.

8 Foundation of a practical theory of project management

Louis Lousberg and Hans Wamelink

Introduction

Literature shows that 'the underlying theory of project management is obsolete' (Koskela and Howell, 2002a), that 'in prior literature it has been generally seen that there is no explicit theory of project management' (Koskela and Howell, 2002a) and that 'several prominent authors have raised the need to introduce alternative theoretical approaches to the study of projects, and to identify the implications that they may have for how we organize and manage projects' (Cicmil and Hodgson, 2006). The quest for a theory of project management can therefore be considered as problematic. From a project manager's point of view it seems to be important that project management theories are practical; practical in the sense that it enhances practitioners' understanding and practical in that it works. The central question that we address in this chapter is thus: what is the foundation of a practical theory of project management?

'The linguistic turn' as a start

In the search for theories that underlie the Project Management Body of Knowledge (PMBOK®) (PMI, 2004), it can be concluded that anomalies occurring in the application of these underlying project management theories are regarded as 'strong enough for the claim that a paradigmatic transformation of the discipline of project management is needed' (Koskela and Howell, 2002a). Here, it is proposed to take the 'linguistic turn' (Rorty, 1967) as the start of that paradigm shift.

> The roots of the linguistic turn lie in a stream of work in philosophy concerned with the nature of meaning and experience. The linguistic turn describes a particular philosophical understanding that proposes 'a particular relation of language to social/historical embedded "seeings" of the world and every person's situated existence' (Deetz, 2003).
>
> (Clegg, 2005)

It is interesting that Deetz's article describes the linguistic turn as one of the historical attempts to escape the subject/object dualism and the assumption of a psychological foundation of experience started by Husserl (1913/1962):

> In his treatment, specific personal experiences and objects of the world are not given in a constant way but are outcomes of a presubjective, pre-objective inseparable relationship between constitutive activities and the 'stuff' being constituted. Thus, the science of objects was enabled by a prior but invisible set of practices that constituted specific objects and presented them as given in nature. And, the presence of personal experiences as psychological, required first a constituting perspective, invisible and prereflective, through which experiences were possible. A floating/social/historical/cultural/intersubjective 'I' thus always preceded either the objects of science or the psychological 'I' of personal experience.
>
> (Deetz, 2003)

Deetz continues:

> Most objects and experiences come to us as a sedimentation from their formative conditions. They are taken as our own or in the world, and the specific conditions of their formation are forgotten. These 'perspectives' or 'standpoints' are institutionalized and embedded in formed experiences and language, and as such, invisibly taken on as one's own, while they are reproductions of experiences originally produced somewhere else by others. These 'positions' or 'standpoints' are unavoidably political.
>
> (Deetz, 2003)

So language no longer represents reality: it is reality itself. Contrary to 'the problem of language as the "mirror of nature" that preoccupied the positivist' (Deetz, 2003) (social) reality is here regarded as a construct.

Theses

Theory and approaches

A striking feature of discussions about theories of project management is their frequent failure to clearly distinguish between such terms as project management, theory, approach and paradigm, or to identify connections between them. This lack of clarity, in turn, has the effect of blurring the overall discussion. With the definition of paradigm[1] by Kuhn (1996) in mind, it can even lead to remarkable statements such as 'lean production can be understood as a new paradigm' (Howell *et al.*, 2004). Project management is here defined as an act and theory as consisting 'primarily from concepts and causal relationships that relate these concepts' (Koskela and Howell, 2002a). It is not, however, always clear whether theory precedes (lies under) this act or is constituted in this act (Glaser and Strauss, 1967). This depends on the approach that is used. From a systems theory approach, other theories about the same subject will evolve than those, for instance, from a social science approach. How different approaches generate different theories is illustrated by the following reflection on success and failure in projects:

> Despite the levels of research founded on the presumptions of instrumental rationality in decision making and control, it is increasingly apparent that accepting and applying such orthodoxy does not eliminate project failures, nor does it guarantee project success. The issue of ambiguity associated with qualifying a project as success or failure has attracted scholarly attention. The debate focuses on a more strategic level of decision making, in which project failure appears to be strategic rather than linked to technical problems, and is seen as a result of political processes of resistance in organizations.
>
> (Cicmil and Hodgson, 2006)

Table 8.1 summarizes approaches to understanding project failure by distinguishing three perspectives and linking them to the wider domain of the project management process.

So, different perspectives or approaches lead to different definitions of success and failure wherein different concepts are used. Questioning

Table 8.1 Perspectives on project success and failure

Perspective	Form of organizational behavior and action	Methodological focus	Success and failure seen as
Rational/ normative	Organizational goals; managerial and organizational structures surrounding the project	Simple cause and effect	Objective and polarized states
Processual	Organizational and sociological processes; projects as a form of decision outcome	Socio-technical interaction	Outcomes of organizational processes
Narrative	Organizational and socio-political processes; symbolic action; themes	Interpretation and sense-making; rhetoric and persuasion; critical/ hermeneutics	Social constructs; paradigms

Source: adapted from Fincham, 2002: in Cicmil and Hodgson, 2006.

success and failure therefore leads to different concepts and causal relationships that relate these concepts, hence different theories. This section therefore concludes with the thesis that there is not one theory, but there are multiple approaches.

Approaches and forms of management

While, as we saw above, taking different approaches can lead to different theories of project management, different approaches can also lead to different forms of management. Or rather, just as a problem can be described using different concepts, normative as problems are (De Leeuw, 2002), the solutions to this problem – that is, forms of management – can be described in different terms. It follows, then, that the form of management that is thought to be suitable will depend on the way in which a problem has been described (Wamelink, 2006). The dominant variable that is used to distinguish between different forms of steering is that of uncertainty/complexity (De Leeuw, 2002). Drawing on De Leeuw's five forms of steering, in this chapter, we identify three forms of managing projects: project management (as defined, for example, in PMBOK®), program management and process management. These can be located on an increasing scale of complexity/uncertainty, stretching from 'routine' at one end, to 'improvisation' at the other (Table 8.2).

In order to refine the analysis, it is necessary to further clarify the difference between project management and process management. In this chapter, we follow the definition of process management as 'managing complexity within people networks' (Teisman, 2001). One application of this, for instance, might be an agreement on the rules that project

Table 8.2 Different forms of managing projects along an axis of increasing complexity

Amount of uncertainty	Form of steering	Form of management	Examples
Very low	Open loop	Routine	Managing industrial fabrication
Average	Feedback	Project management	Managing systems
Reasonable	Feed forward	Program management	Managing policy
High (also ambiguity)	Meta	Process management	Managing interaction
Very high (also ambiguity)	Intrinsic	Improvization	Managing brainstorms

participants will follow in order to reach a decision. Another definition of process management is that adopted by Bekkering *et al.*: management of the development of ideas (Bekkering *et al.*, 2004). Again, this definition is not about content per se (as with the realization of a preconceived idea, for example), but merely about the process of getting to an idea. In the literature, the concept of process management is presented in opposition to that of project management and Table 8.3 offers an example of this tendency.

While taking such an approach certainly clarifies the differences between the two, as suggested above, several studies have emphasized the differences between project/systems management and process/interaction management, and come out in favor of the latter. Literature thus suggests that even within one project, both approaches can be valuable, depending upon the issue at hand (Groote *et al.*, 2002; Bekkering *et al.*, 2004). As experienced project managers, the authors of this chapter fully agree with this point; as a project manager, one has to be able to shift quickly from taking a project approach to taking a process approach. So, different forms of management are suitable for different problems, questions and subjects; and all of these different forms of management can be appropriate within the context of a single project. Hence, we can conclude this section with the thesis that there is not one form of managing projects, but there are several.

Knowledge gap and interpretative research

Although different problems or questions can be addressed by using different forms of project management, the existing literature on project management is dominated by a discourse in which the instrumental project approach and form outweighs that of the social process approach

Table 8.3 Differences between project and process management (after Teisman, 2001)

Project	Process
One time activity	Multiple activity
One goal	Several goals
Limited time	Long time orientation
Heterogeneous in pattern of action	Heterogeneous, ambiguous and
Temporary organization	dynamic
Uncertainty	Organization of interaction
Production out of line management	Uncertainty and ambiguity
Violates well-known conventions	Production in arenas within
Disturbs line organizations	organizations
	Seeks new conventions
	Generates dynamics and requires
	flexibility

and form (Lousberg, 2006; Howell *et al.*, 2004; Cicmil and Hodgson, 2006; Cicmil *et al.*, 2006). Several attempts have been made to rectify this imbalance; but once again, potential solutions have taken the form of attempts to apply instruments, such as in the case of *Last Planner* and *Scrum* software (Koskela and Howell, 2002b; Ballard, 1994; Howell *et al.*, 2004; Aravena, 2005). While it has been acknowledged that current project management fails to create the conversations necessary to develop a shared background of obviousness and common concerns (Howell *et al.*, 2004), this acknowledgement does not seem to be based on an under-standing – that is, an internalized understanding, or *verstehen*, as Max Weber put it – of the social processes occurring within a project that might lead to alternative approaches. Future research programs should therefore focus on this gap in our knowledge of the social aspects of project management.

Just as there is not one theory of project management, in the social sciences we seem to have moved beyond the era of 'grand theory'; that is, the notion that theory might provide pre-existing and universal explanations for social behavior. From a social science perspective, contemporary project management research is, in part, focused on specific and context dependent phenomena. In the following paragraphs, we offer three examples of this type of research.

In their study, 'Governmentality matters: designing an alliance culture of inter-organizational collaboration for managing projects', Clegg *et al.* (2002) investigated how a project to design and build a sewage facility in Sydney Harbour was successfully completed, on time and within budget, prior to the start of the 2002 Olympic Games. The project commenced with an alliance contract that contained a minimal number of require-ments. The project's strategy, specifications and design had to be developed by means of interaction within the project team, and by fine-tuning the approach to the project environment; that is, by talking.

The key theme for the analysis of Clegg *et al.* became the project culture and its relationship to a set of key performance indicators (KPIs) of schedule, budget, occupational health and safety, community and ecology. Extensive research was undertaken, based on written texts, arti-facts such as posters, and the records of meetings (for instance, over 1,000 pages of transcripts were analyzed). This led to the finding that *govern-mentality* poses an alternative to policing, litigation and arbitration, espe-cially in situations of multiple actors and interests, through the design of a more collective and coherent practical consciousness within which to make sense.

This example clearly shows how the instrumental can play an important role in the management of a project. Furthermore, parallels can be drawn with the aim of Howell *et al.* to design a more collective and coherent practical consciousness within which to make sense and to develop a shared background of obviousness and common concerns (Howell *et al.*,

2004). Unlike software, however, the instrument is not invented and then implemented. Rather, it is invented in the process of talking and thus emerges from the challenges that have to be met in the course of managing the project. This probably leads to far more effective solutions than mere software implementation, although the latter approach, of course, can also be useful.

A second example is that of Cicmil's study, *An Inquiry into Project Managers and Skills* (Cicmil, 2006). In order to answer the question, what it might mean and take to be a high performer or a virtuoso project manager?, interviewees were asked to reflect in an open-ended way on such themes as key challenges, their own performance, their personal careers and the role of training. Of particular interest is the fact that the authors described their research methodology as originating from a pragmatic epistemology, designed as a participative cooperative inquiry based on active interviewing, involving reflective practitioners and pragmatic researchers. Some of the insights into project management practice that emerged from this cooperative inquiry include continuous renegotiation of the project's direction and plans, experienced in a social context where conversations and power play an equally important role as documents and procedures; and understanding project management as a social and political action in context: evaluating the situation using judgment, intuition, previous experience and a holistic, multi-perspective approach as well as logic and universal principles of project management to act and perform in the specific local context of the living present.

These findings confirm the point made earlier in this chapter regarding the supposed coexistence of instrumental and social forms of project management. This example also suggests that the role of project managers as implementers can be a problematic one. Most importantly, however, it illustrates that taking practice as the basis for research can reveal a different vision of everyday project management from that more commonly provided in the project management research literature, and the literature is further enriched as a result.

A final example of research that focuses on the specific, context dependent aspects of project management consists of part of one of the author's doctoral research into conflicts in complex public–private spatial planning projects. This research examines how conflicts might be managed to avoid them becoming dysfunctional. Part of the research consists of case studies, which were analyzed to obtain insights into the evolution, and possibly the causes, of conflicts in specific contexts. The existing literature suggests that differences in perception play an important role in the emergence of conflicts. To confront this theory with practice, a method was chosen that analyses the actual production of meanings and concepts used by social actors in real settings (Suddaby, 2006). The purpose of the analysis was to make statements about how actors interpret reality, rather than obtaining scientific truths that are based on 'reality'.

For this research, transcriptions were made of open-ended interviews, in which conflicts, dysfunctional conflicts and solutions were discussed. Next, theoretical concepts and relations between these concepts were interpreted but only if grounded in the raw data.

The findings were as follows:

- in the conflict case, the dominant factor in the escalation into a dysfunctional conflict seemed to be a mismatch between the images that the architect and the project developer had of one other; and
- in the non-conflict case, there seemed to be a relationship between preventing conflicts and understanding differences in areas such as quality, costs and revenues (in short, economic feasibility).

These findings then provided the basis for a hypothesis on how to prevent dysfunctional conflicts.

This third example thus illustrates that, far from being derived from a theory that had been used to guide data collection and analysis, concepts and the relationships between them emerge from data and the subsequent analysis of data (Suddaby, 2006).

What these three examples of specific, context dependent research have in common is that they:

- make project management practice the basis of the research;
- use interpretative qualitative research methods;
- investigate the construction of a shared reality and of how theory can be shaped by reality;
- distinguish between the instrumental/rational and social/personal, while studying both; and
- deliver deep insights into the practice of project management.

Therefore we conclude this section with the thesis that the current imbalance between knowledge about project management's instrumental and social forms can be rectified, in favor of the latter, by research that focuses on specific, context dependent practice that is grounded in what practitioners say about practice.

Conclusions

The central question addressed in this chapter is: what is the foundation of a practical theory of project management? That is, practical in the sense that it enhances practitioners' understanding of project management and practical in the sense that it works.

In the course of the discussion, three theses were elaborated:

1   there is not one theory of project management, but there are in fact multiple approaches;

2   there is not one form of managing projects, but there are several; and

3   due to the gap of knowledge of social project management forms, research should be focused on specific, context dependent practice and be grounded in what practitioners say about this practice.

From these three theses, it is concluded that the foundation of a practical theory of project management that enhances practitioners' understanding of project management, and that seems to work, is practice itself.

Note

1 A commonly accepted scientific achievement that delivers for a certain time model problems and solutions to a community of researchers (Kuhn, 1996).

9 Role of action research in dealing with a traditional process

Seirgei Miller, Henny ter Huerne and André Dorée

Introduction

Over the last few years, since a parliamentary enquiry into the construction industry, the business environment in the Netherlands has changed dramatically, not least in the road construction sector. According to Dorée (2004), the collusion structure that regulated competition has fallen apart. Public clients have introduced new contracting schemes containing incentives for better quality of work (Sijpersma and Buur, 2005). These new types of contracts, tougher competition and the urge to make a distinction in the market, have spurred companies to press ahead with product and process improvement. These changes have significantly altered the playing field for competition. The companies see themselves confronted with new 'rules of the game'. Performance contracting and longer guarantee periods have created different sets of risk, but also incentives. In general, companies are experiencing pressure from new types of competition and other rules and trends, but at the same time they recognize the need and the opportunity to differentiate themselves.

The road construction sector offers valuable insights into how this new order is shaping an industry that has been used to longstanding ways of working. This chapter examines one important aspect of road construction – the asphalt paving process. By focusing on a well defined and economically significant area, it is possible to mount an in-depth study with reasonable expectation of being able to take and apply the methodology and insights to other sectors of construction. In an effort to outperform competitors, asphalt paving companies are seeking better control over the paving process, over the planning and scheduling of resources and work, and over performance. Improved control can also reduce the risks of failure of the paving during the period of guarantee. To be able to achieve these goals, the relevant operational parameters need to be known and the relationships between these parameters have to be thoroughly understood. For asphalt paving companies to be able to improve both product and process performance, they need to develop a more sophisticated understanding of the asphalt paving process and the interdependencies within it.

The research reported here forms part of a larger project focusing on the improvement of the asphalt paving process aimed at improving quality and reducing its variability. This chapter presents the development of a research strategy to address two key research questions. The first question relates to the main causes of variability in the asphalt paving process and the second is concerned with the effect of revised operational strategies on quality in this process.

Methodological considerations

During a workshop conducted by Dorée and ter Huerne (2005), national experts and representatives of agencies in the asphalt paving field were confronted about the state of road construction in the Netherlands. The experts suggested that:

- little or no research effort is put into the systematic analysis and mapping of the asphalt paving process;
- the process depends heavily on craftsmanship;
- work is carried out without the instruments to monitor key process parameters; and
- selection of work methods and equipment is based on tradition and custom.

We undertook three tasks in response to anecdotal suggestions made during the workshop. First, we conducted an extensive literature review to assess the state of research into the asphalt paving process. Second, we conducted one-on-one on-site interviews with 28 machine operators. The purpose was to gain insight into operational strategies in the process from the perspective of the operators and thereby confront the suggestions made by the national experts. Last, we developed a research strategy to move the process forward in an attempt to answer the research questions mentioned above.

State-of-the-art review

Research into the asphalt paving process

Several agencies and organizations dedicated to asphalt research exist in the Netherlands and abroad. A scan of literature on asphalt issues showed a field of asphalt research that is well developed. One area dominates the core of our knowledge base to date – the characteristics of asphalt as a construction material (e.g. mixtures, recipes, strengths and elasticity). Efforts to systematically map and analyze the process of asphalt paving are, however, comparatively few. Approximately 100 papers were published in the *International Journal of Pavement Engineering* during the period 2002 to 2005 of which just one was in the construction process research area. A similar situation applies to the *International Journal of Pavements* during that same period, with a mere two papers out of 65 (approximately 3 percent) directly addressing construction modeling. A scan of publications in the *Journal of Computing in Civil Engineering* revealed that six papers (approximately 5 percent) were published in the areas of modeling and simulation of construction processes.

Abudayyeh *et al.* (2004) investigated construction research trends in technical papers published in the American Society of Civil Engineering's *Journal of Construction Engineering and Management* between 1985 and 2002. In all, 879 technical papers were analyzed. The top research areas were reported as scheduling, productivity, constructability, simulation and cost control. These topics formed approximately 18 percent of the total number of papers published during that period. It is interesting to note that the modeling of construction processes comprised less than 2 percent of the total number of papers published during this period and that it ranked a mere 17th out of a list of 29 research areas. Despite the apparent neglect of construction process research, a positive trend appeared in the period 1997 to 2006 with an increase in the number of construction simulation papers (see, for example, Sawhney *et al.*, 1998; Halpin and Martinez, 1999; Naresh and Jahren, 1999; Kartam and Flood, 2000; Halpin and Kueckmann, 2002). This trend continued after 2002 (Zayed and Halpin, 2004; Zhang and Tam, 2005) albeit with few papers covering simulation of the asphalt paving process (White *et al.*, 2002; Jiang, 2003; Nassar *et al.*, 2005; Choi and Minchin, 2006).

We can therefore conclude that the majority of the research and papers deal with the characteristics of asphalt as a construction material. Research into the asphalt paving process is in its infancy.

Mapping the asphalt paving process

There have been several organized industry supported research efforts for the development of state-of-the-art technology for real-time locating and positioning systems for construction operations (AbouRizk and Shi, 1994; Pampagnin *et al.*, 1998; Bouvet *et al.*, 2001; Hildreth *et al.*, 2005; Navon *et al.*, 2004). They include efforts to develop automated methods for monitoring asphalt laying and compaction using GPS and other ICT.

Li *et al.* (1996) reported on a system for mapping moving compaction equipment and transforming the result into geometrical representations, and investigated the use of GIS technology to develop a graphical representation of the number of compactor passes. Krishnamurthy *et al.* (1998) developed an Automated Paving System (AUTOPAVE) for asphalt paving compaction operations. Peyret *et al.* (2000) have reported on their Computer Integrated Road Construction (CIRC) project. This aims to develop computer integrated construction systems for real time control and monitoring of work performed by road construction equipment, namely compactors (CIRCOM) and pavers (CIRPAV). Oloufa (2002) described the development of a GPS-based automated quality control system for tracking pavement compaction. The Compaction Tracking System (CTS) allows tracking of multiple compactors.

Several experiments to map the asphalt paving experience have been conducted in recent years. While some of these experiments were developed into industrial applications, it appears that few have been accepted widely by industry or used regularly on construction sites. Although some equipment manufacturers now provide GPS as an option for clients, it is not yet part of operational strategies and working practice in asphalt paving processes. For example, GPS technology has been the subject of prior study and, subsequently, deployed on roller equipment; even so, it is not being adopted and integrated into operational strategies and methods.

Results from on-site interviews

The interviews revealed several tensions between theory and practice. A major practical problem which roller operators have to resolve is an inability to measure the degree of compaction during the compaction process despite being responsible for the final compacted level of the asphalt mat. When final rolling has stopped the target density should ideally have been achieved since it would be difficult to achieve further compaction once the asphalt mat has cooled down (Timm *et al.*, 2001).

Most of the operators interviewed indicated that they were not informed of the final density of the completed layer – not during the site operations or even afterwards – in spite of its importance. This is a significant shortcoming in terms of quality control. It shows an absence of 'closing the feedback loop' (Montgomery, 2005) and as such negatively affects any learning that could occur. Furthermore, the number of roller passes and roller patterns directly influence compaction (Leech and Powell, 1974; Roberts *et al.*, 1991). While indicating that they used prescribed roller patterns during compaction, one concern is that most operators did not keep track of the number of passes completed during rolling. They also appear to base key operational decisions on what they feel and see, since they do not know what the actual temperatures and the material characteristics are during compaction.

Roller operators indicated that they specifically looked for the occurrence of cracking and shoving, as well as the rapid cooling of the asphalt during compaction. Interestingly, the speed of the asphalt paver was not considered an important enough factor to warrant discussion between screed and roller operators. This raises the issue of whether or not they were aware of the effect of temperature differentials if the paver was too far away from the roller's working zone. The influence of temperature differentials on hot mix asphalt paving has been studied extensively (Chadboum *et al.*, 1998; Timm *et al.*, 2001; Stroup-Gardiner *et al.*, 2002;

Willoughby *et al.*, 2002). The relevance of temperature issues seems in stark contrast with the road crews' (lack of) attention to this parameter during the paving process.

Evidence of barriers to technology adoption was revealed in a number of ways. Most operators frankly acknowledged that they hardly made use of the technology available on their machines. Of the operators that had temperature measurement tools at their disposal, only a minority confirmed their use of them. They showed an awareness of the importance of the cooling process (of the asphalt) and considered weather conditions, temperature of the asphalt mix and changes in layer thickness to be factors important enough to warrant their attention during paving. They also understood that a change in layer thickness directly affects the cooling rate of the asphalt mat. It is easier to achieve target density in thicker layers of asphalt than in thinner layers. This is because the thicker the mat, the longer it retains the heat and, thus, the more time there is available to achieve compaction (Asphalt Institute, 2007). Even so, in practice the roller operators are mostly uninformed about such discontinuities because of adjustments made by the paver and screed operators, for example paver speed, layer thickness and screed vibration.

The operators confirmed anecdotal evidence which suggested that, in the Netherlands, work in the asphalt paving process depends heavily on craftsmanship, that work is being carried out without measuring key process parameters (temperature, density and layer thickness) and that work methods and equipment are selected on the basis of tradition and custom. There is also evidence that no direct feedback is given to machine operators. Machine settings are done mainly on the basis of feel and experience. Although the interviewees referred to common and proven practice in machine setting, actual settings and operational strategies varied widely from team to team.

Asphalt paving in many ways is still a process driven by craftsmanship, heavily dependent on tradition, and on operators' experience, gut feeling and tacit knowledge. There is not really one common practice, but a wide range of common practices that leads to extensive variability in the quality of the final product.

Action research strategy

Given that craftsmanship still rules operational choices in the paving process and that operational strategies are typically tacit, can new technology provide an impetus toward a more professional approach? This is not a straightforward 'yes' as often assumed. Our interviews indicated that operators are not comfortable with new technology. Over the last ten years new technology has been developed to improve process information and process control (see the previous section). New features and functions have been added to equipment. Most operators acknowledged that they hardly made use of available technology. They just do not know how or when to use the new gadgets.

Another relatively new technology is GPS. Although equipment manufacturers now provide GPS as an option, it is not yet adopted and integrated into operational strategies and methods. In fact, the data provided by GPS systems does not help operators, because they do not know if these data might be relevant for their operational choices and work methods.

The adoption of technology process may also be hindered by skepticism and reluctance on the part of operators who feel that their workmanship is being devalued or that management could use the technology to track their movements and possibly use it punitively (Simons, 2006). Several authors argue that the construction industry typically lags behind other industries in adopting new technology (AbouRizk *et al.*, 1992; Halpin and Martinez, 1999; Halpin and Kueckmann, 2002; Bowden *et al.*, 2006).

Evaluating the adoption of new technology can be accomplished by using the innovation adoption factors determined by Rogers (2003), namely relative advantage, compatibly, complexity, *trialability* and *observability*. When data produced by the GPS systems do not match operators' operational reasoning, at least three of Rogers' five attributes will not be fulfilled and adoption will be problematic. At the same time, tailoring GPS solutions to overcome this mismatch is difficult because the operational reasoning of the operators is tacit and implicit.

New research approach

Developing better operational strategies requires that new technology be adopted, yet it is not because of insufficient understanding of current operational strategies (the 'common practice'). This resembles a 'chicken or egg' problem, a causality dilemma. Against that background, the research project has followed an action research strategy alternating steps of technology introduction with the mapping of operational strategies (see Figure 9.1). Through monitoring of the learning processes of the operators, and evaluating the operational choices with them, the tacit knowledge of 'common practice' will become explicit. This provides the opening for further development of process understanding, tools and operational strategies. Qualitative heuristics will be confronted with quantitative process data.

The proposed strategy implies that, first, it involves the asphalt machine operators directly in the research project and, second, it includes a statistical modeling and computer simulation component that aims to test and validate models developed during the research. The explicit models will facilitate the practitioners in synthesizing their tacit knowledge and promoting learning processes. Trochim (2001) suggests that:

> there is much value in mixing quantitative and qualitative research. Quantitative research excels at summarizing large amounts of data and

Figure 9.1 Action research strategy (adapted from Susman, 1983).

reaching generalizations based on statistical projections. Qualitative research excels at telling the story from the participant's viewpoint, providing the rich descriptive detail that sets quantitative results into their human context.

The aim is for operators and researchers to develop joint operational strategies using an iterative process (see Figures 9.2 and 9.3) of problem definition, operational strategy development, implementation, evaluation and consciously specifying the learning taking place. This is expected to lead to:

• better understanding of the asphalt paving process;
• the development of innovative tools and technology to assist understanding of the process; and
• adoption and wider acceptance of innovative tools and technology and their associated benefits.

A qualitative paradigm should provide insight and understanding from the perspective of those actually involved in the asphalt paving process. One of the major distinguishing characteristics of qualitative research is that the researcher attempts to understand people in terms of their own definition of the world (Mouton, 2001). By utilizing a qualitative approach, an attempt will be made to understand the asphalt paving process, from the subjective perspective of the individuals involved. These individuals include the operators in the paving process. The complexities

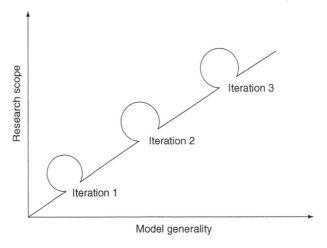

Figure 9.2 Iteration process (adapted from Kock *et al.*, 2000).

can only be captured by describing what really happens when they are doing their job, incorporating the context in which they operate, as well as their frame of reference. In other words, there needs to be a commitment to the empowerment of participants and the transfer of knowledge. Chisholm and Elden (1993) advise that one should strive for the full involvement of the client (in this case the machine operators) and the researcher. The involvement of participants enhances the chances of high construct validity, low refusal rates and ownership of findings. The validity should also benefit from several iterations and expansion of the research scope across those iterations (Kock *et al.*, 2000). This is shown in Figure 9.2.

A qualitative approach therefore has the potential to supplement and reorient our current understanding of the asphalt paving process. Key research questions using an action research strategy are normally of an exploratory and descriptive nature. Exploratory in that you are attempting to, first, assess what is happening during the process and, second, to

Figure 9.3 Typical iteration (adapted from Kock *et al.*, 2000).

identify the key factors that affect that process. Descriptive questions also provide opportunities for finding correlations between variables affecting the process.

The quantitative paradigm is aimed at developing and validating accurate models of a somewhat complex process. The overall objective is to build models of the process and to bring them together in an event scheduling system. The models to be developed need to be checked and validated in practice. This requires the involvement of stakeholders closest to the asphalt paving process. Several causal and predictive questions have to be addressed during this modeling phase. What are the main causes of variability in the process? Is variability the main cause of reduced quality, productivity and efficiency within the process? What will be the effect of a revised operational strategy on the process? Will a revised operational strategy lead to improved quality, productivity and efficiency?

With this action research strategy, the 'chicken or egg' problem (the causality dilemma of technology development and adoption) is side-stepped by progressing incrementally through the agency of practitioners. The action research strategy described here has an added benefit. Since progress in the research project coincides with actual learning and growth of operational knowledge and capabilities, the companies are happy to take part, instead of merely being the object of study. It breaches the divide between science and practice. It not only challenges practitioners' presumptions about the paving process, but also their opinions of the value of researchers and research.

Conclusions

A parliamentary inquiry into collusion in the Dutch construction industry has sparked new public procurement strategies and altered the business environment for road paving companies. Performance contracting and extended guarantee periods now drive the companies toward improvement in product quality and process control. Since the density of a pavement is a key factor in the strength and durability of the surface, operational strategies are a cardinal focus for research. Attention to these issues has revealed that site operations and operational strategies are driven by 'common practice' – the tacit knowledge and heuristics of the site crew built on years of personal experience (and often idiosyncrasies). Building an objective picture of site operations is difficult since they are not documented. Knowing the exact location of construction vehicles, their speed and motion characteristics, can provide essential information for better understanding the asphalt paving process. This can be done using GPS technology, but it is not straightforward. Experiments with such technology introduction show problems in adoption. In order to be adopted, the technology should be tailored to the prevailing operational strategies, but at the same time the technology has to be adapted to make the prevailing operational strategies tangible. To overcome this causal dilemma we propose an action research approach.

This action research approach provides opportunities for developing a framework to capture the operational characteristics of the asphalt paving process in a more holistic manner. It diverts from previous process modeling studies where key role players have been left out of the process. Latham (as cited in Blockley and Godfrey, 2000) observed that 'there is an acceptance that a greater interdisciplinary approach is necessary, without losing the expertise of individual professions'. He recognized that all concerned with construction are interdependent and need to behave as a team. Blockley and Godfrey (2000) also argue that 'we need to have a whole new view of process' and in order 'to do that we need to include factors that are particularly needed when co-operation between people is important'. The key issue here is that operators need to be involved in, and take responsibility for, the process. They are, in fact, largely responsible for the success of the process.

The action research methodology involves the researcher, innovative technology and, most importantly, the machine operators in 'driving' the asphalt paving process. The first steps in this project show that the approach taps into the enormous wealth of tacit knowledge and experience of operators – it provides insights necessary for analyzing important operational characteristics in the process. Unraveling and confronting the practitioners' view is expected to lead to improved control during the process and, consequently, to improved product and process performance.

Acknowledgment

The authors wish to thank Bart Simons for conducting the interviews with the on-site operators.

10 Corporate strategies

For whom and for what?

Johan Björnström, Ann-Charlotte Stenberg and Christine Räisänen

Introduction

Effective strategic management is becoming an increasingly important issue both for practitioners and management scholars. Not only is the process of formulating and implementing strategies given higher priority, but the role and meaning of strategies are also changing (Price, 2003). In the construction industry, however, relatively few companies seem, as yet, to have established a formal strategy process, even though there is considered to be greater awareness of the importance of effective strategic management to enhance performance and profitability (Junnonen, 1998). In the purportedly conservative construction industry, actors prefer adhering to the 'business as usual' mindset, which often results in a drift of strategic meanings and ultimate blurring of the organization's strategic position (Johnson *et al.*, 2005).

Following a number of reports of companies' failure to implement strategies (see, for example, Allio, 2005; Corboy and O'Corrbui, 1999; Kaplan and Norton, 2001), the attention of practitioners and researchers is now shifting from the formulation process to implementation dilemmas (Aaltonen and Ikavalko, 2002). The already growing body of research into strategy implementation seems to agree that one of the main reasons for failure is ineffective organizational communication caused by a lack of consideration of the social environment at the strategy execution level of the organization (Miniace and Falter, 1996). Yet, what is meant by the term 'communication' is not defined, and just a few studies have focused on the discursive and rhetorical aspects of strategy communication (Fairhurst *et al.*, 1997; Johansson, 2003; Müllern and Stein, 1999). These studies typically describe managerial strategic communication as being transactional rather than interactional, monologic rather than dialogic and top-down rather than bottom-up. They also characterize strategic rhetoric at the top level of management as abstract rather than concrete, idealistic rather than realistic and distanced rather than proximal.

To our knowledge, no such studies have been carried out in the construction industry. The overall purpose of this chapter is, therefore, to report preliminary results from a longitudinal case study of the strategy work carried out in a large Swedish construction company during a period of organizational change. Our concern here is the ways in which the new strategies are communicated down the chain of command in the company: from top management levels via middle management to project management. We focus on the face-to-face communications used by the different managerial levels to disseminate the corporate strategy and the implications this has on the ways in which the strategies are interpreted and understood. Of particular interest in these interactions are the underlying reasons for the different approaches toward strategy implementation. We hope to contribute some insights into the complexity of communicative processes and practices and argue that organizations need to view discursive processes and practices as an integral part of organizing.

Frame of reference

The word 'strategy' gives rise to a multitude of meanings and definitions depending on the perspectives of the respective writers. It is a term that has been widely treated throughout history, giving rise to different, and often conflicting, views (Price, 2003). Scholars have tried to classify the different views or schools of thought concerning strategies, for example Bailey and Johnson (1992), Chaffee (1985), Whittington (1993) and Mintzberg *et al.* (1998). As in all such classifications, there is a fair amount of overlap between the categories (Faulkner and Campbell, 2003).

Here, we discuss Whittington's (1993) classification of strategy into four perspectives: *classicist, evolutionary, processual* and *systemic*. Depending on the perspective or approach toward strategy management, the processes through which the strategies are generated and the implications for the outcomes will differ (Figure 10.1). We will concentrate on three of these perspectives: *classicist, processual* and *systemic*.

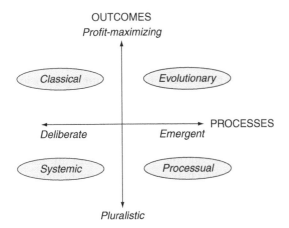

Figure 10.1 Generic perspectives on strategy (after Whittington, 1993: 3).

The *classical* perspective originated in eighteenth-century economics, drawing on the militaristic ideals of Ancient Greece. For *classicists*, profitability is the ultimate business goal and rational planning is the only means of attaining it. The classical approach in business only emerged as a coherent discipline in the 1960s (Whittington, 1993) with the idealization of the CEO who commands his people to carry out his formulated strategies without question.

The *processual* perspective views strategies as emergent, generated in response to the chaotic environment of organizations and markets. Cyert and March (1963) claim that people are only 'boundedly rational' which they explain as follows:

> [W]e are unable to consider more than a handful of factors at a time; we are reluctant to embark on unlimited searches for relevant information; we are biased in our interpretation of data; and finally we are prone to accept the first satisfactory option that presents itself, rather than insisting on the best.
>
> (Cyert and March (1963) in Whittington (1993: 23))

Furthermore, the *processual* perspective sees organizations as made up of individuals, who bring their own needs, wishes and biases to their places of work. To work toward an organizational vision and goal in such heterogeneous groupings requires negotiation. The consequences are most often that several goals will apply and these may not always be aligned with each other. Negotiations combined with bounded rationality may, according to Whittington, lead to strategic conservatism and resistance to change. Moreover, in this perspective, strategies emerge retrospectively and are seen as a management tool to describe and simplify a chaotic world.

The *systemic* perspective sees organizational behavior as deeply rooted in employees' social networks and strongly influenced by family, education, background, religion and ethnicity (Swedberg *et al.*, 1987; Whittington, 1993). Thus, the strategies reflect the social systems in which they are formulated and enacted. The concept of formally planned strategies is a culturally constructed phenomenon, predominantly advocated in Anglo-Saxon culture. From the systemic perspective, formal planning is sociologically efficient although maybe not economically so. Since formal planning does satisfy a fundamental human need to order one's environment, it is therefore beneficial to organizations.

These perspectives on organizational strategies entail different approaches to strategy formulation and implementation. How strategies are used and understood at various managerial levels may be very different from how they are interpreted and understood at the operational level. Consequently, these perspectives presuppose very different communicative approaches and discourses. The classical perspective assumes a top-down, one-way, transmission approach in which strategies are seen as information or orders to be carried out by middle management and down the chain of command. In the *systemic* and *processual* perspectives, strategies are seen as constructed entities that are negotiated by heterogeneous groups and may be enacted, but may also give rise to resistance and non-action. With these views, communication becomes a means for reaching a consensus and shared knowledge. This would assume a two-way, interactive communication.

As mentioned earlier, managers seem to lack communicative competence (Johansson, 2003). This appears to apply in the construction industry too, which has been criticized for its patriarchal and conservative mindset (Knauseder, 2007; Gluch, 2005). In our view, this lacuna cannot be bridged by simply attending prescriptive 'how-to' communication courses. What is needed is a theoretical understanding of the cognitive processes that take place over a chain of communicative acts where ideas or propositions are conveyed and interpreted. One explanatory lens is that provided by Czarniawska and Joerges (1996), who see ideas (be they in non-materialized forms such as thoughts and talk or materialized in artifacts such as textual inscriptions) as traveling through different times and spaces. As the ideas move between humans or from humans to artifacts and back again, they take on altered states as they collide with competing ideas.

Traditionally, innovations and new ideas have been viewed as being spread through a diffusion process (Rogers, 2003). The weakness of the diffusion model is that the recipient of an idea is regarded as a willing receiver, whose slate is empty. Diffusion is therefore a model of a linear, one-way communicative process (Powell and DiMaggio, 1991). Moreover, the diffusion model also presupposes that ideas do not alter as they are transferred from individual to individual or group to group. In contrast to the diffusion model, Czarniawska and Joerges (1996), following Latour (1991), show how ideas go through chains of translations, namely they alter as they move from sender to receiver in order to suit new contexts of use and new users (Meyer, 1996).

One limitation of the 'travels of ideas' metaphor is that it does not contribute to a full understanding of *what* happens and *why* when an idea surfaces and is caught by actors in a particular local space. It shows *that* ideas travel and change, but not *how* this change takes place or what factors may contribute to the change on organizational, group or individual levels.

Research method

The empirical analysis for the research presented here is based on a longitudinal case study (Yin, 2003), carried out between 2005 and 2007, on strategy implementation in a construction company during a period of change. The case organization, hereafter called *Constructo*, is a large global construction corporation as well as the largest domestic contractor with approximately 12,000 employees. The case study covers the domestic arm only. The new organizational pyramid consisted of the top management and its operational development team, 26 regions and 107 districts. For the purpose of this chapter, the strategic work in one region and its four districts were studied.

Three iterative approaches were used to collect the data for this part of the study. The first approach consisted of 13 semi-structured interviews of which two were follow-up interviews. The group of interviewees consisted of one regional manager, one regional staff manager, four district managers, one project manager and four production managers. The interviews lasted between 45 and 105 minutes and were audio recorded then transcribed in full. The second approach consisted of field observations, where the researchers participated in seven meetings on strategy work in one business unit, hereafter referred to as Region A. All the meetings were audio recorded and field notes were taken. After each meeting, the data were reviewed and analyzed by the authors. The third approach was an analysis of documents, business plans, slide presentations and personal notes as well as follow-up documentation collected from participants.

Different approaches to corporate strategies

At the time of the study, the organizational structure at *Constructo* consisted of a top management team supported by an operational development team. Below top management, the organization was divided into functional and geographical regions, e.g. housing, road construction and civil engineering. The regions were divided into districts, essentially operational units that undertook the business in local markets. The function of the regions was to manage, support and coordinate bundles of four to eight districts.

The different management groups at *Constructo* were formed so that there would be a certain overlap of functions. Thus the district managers were part of the regional management group and the project managers were part of the district management group (Figure 10.2). Accordingly, each manager attended strategy meetings at two organizational levels.

Intended work process

Top management reviewed the strategies on an annual basis and aligned them with company operations. Although the strategies applied to the whole

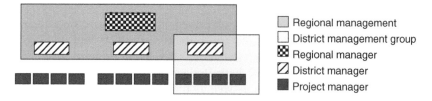

Figure 10.2 The management group affiliation for managers at *Constructo*.

organization, they were mainly formulated to address regional managers, who received the strategy document by email attachment. The work process advocated by top management was that the regional management groups should review the strategies and prioritize according to their own local contexts. Moreover, they were required to formulate action points for fulfilling the strategies. The idea was that the resulting regional business plans would be sent to top management for review and approval and then to the district managers for further processing. The districts in turn were to focus on the execution of the strategies by designing district specific action points.

The key question when prioritizing the strategies was supposed to be 'Do we currently work in accordance with this strategy today?' and not 'Is this strategy important?' In other words, top management gave regional managers some freedom to choose the strategies that applied particularly to their line of work although it was made clear that all the strategies were important from the company's perspective:

> The idea is for each region to review the strategies and ascertain how well they fulfill each one. If performance is not up to the region's desired level, then [that strategy] should be prioritized and something should be done. If [the region] is already satisfied with the work done in another strategic area, it does not have to work so much with the particular strategy at that time. But [the strategy] remains important.
> (Member of the operational development team)

Top management emphasized that the regions should prioritize among the strategies according to their current status. However, we found that managers still prioritized the strategies according to their perceived importance for the region, which was contrary to top management's implicit intentions: 'I have heard people say that "No, we did not choose that strategy. It was not important for us". But that is not the intention and it is serious if it is perceived as such' (Member of the operational development team). From the documents we collected and our interviews, we witnessed a prevalent worry among managers concerning employee engagement and commitment to the strategies. The means for instilling these attitudes in lower level managers and employees was neither discussed nor regarded as an issue. The attitude at the top management level was that the need for improvement should spur the necessary engagement with respect to the strategies and generate action points.

Strategy work at the lower levels, district and project levels, was relatively free from top management intervention; but since there was very little feedback on the resulting business plans and action points, the freedom risked being interpreted as lack of interest. What senior and top management were mainly interested in was the overall company progression. If a region was underachieving, top management wanted this to be made visible so that they could intervene and wield control.

Strategy work at the regional level

At the regional level the strategy management process adhered mostly to the intentions of top management. The regional management group was called to a two-day, off-site workshop to discuss the new formulations of the previous year's strategies. At the start of the meeting the regional manager gave his interpretation of the purpose of the strategies and of the meeting.

> [The strategies are originally formulated] by our top management. They have looked at how we should work with certain issues in *Constructo*. They look at these issues from a top management perspective and now it is our task to establish what we should do in our region in order to work in the same direction.
>
> (Regional manager)

Each strategy was then discussed in the group to ascertain the status of the region vis-à-vis the strategy. As the meeting progressed, the discussion started to blur when the good of the company, as reflected in the strategy formulations, and the good of the region, as experienced by individual members of the group, were in conflict. For example, as a response to a complaint made by one district manager, who announced that he felt that they were doing this work for somebody else and not for the benefit of their unit, the regional manager countered by arguing that their job was to decide what was meaningful for the region as a whole. As a result of this exchange, the regional manager asked the rest of the group to brainstorm what they considered the five most important tasks for the region for the following year. These lists were then briefly checked off to ensure that all the tasks would be included in the business plan. However, some tasks were difficult to include, such as 'having fun at work' since the group could not find a suitable strategy among those advocated by top management.

The regional management group prioritized 18 out of the 30 strategies that were proposed by top management and action points were formulated for each. After this business plan was approved by top management, it was the districts' turn to prioritize and formulate their action points.

Strategy work at the district level

All the management groups at the district level within Region A used face-to-face meetings, as advised, to discuss the strategies and the regional business plan. While most of the districts mainly discussed only those strategies prioritized at the regional level, one district discussed all 30 strategies. The rationale for this procedure, according to the district manager, was that in the regional business plan the strategies had been prioritized from a regional perspective; in the district, however, different strategies may need to be prioritized to suit the district's local context. The result was that this particular district prioritized 16 of the 30 strategies. Out of the 18 strategies prioritized at the regional level, 12 were kept and four were abandoned. It is interesting to note here that communication in these types of groups tended to be a dialogic interplay between the parties present and the participants did not hesitate to ask questions and even contradict their manager. Not all the groups at the district level were, however, equally interactive.

At the district level, we also saw a shift in approach to strategy work compared to the top managerial levels. Here the key question became: 'Is this strategy something that we should be working on?' The districts put more focus on the perceived importance of the strategy from the district perspective, that is, the district's business plan had to fit the current business and lead primarily to profitability for the district. Here, focus was put on trying to engage project managers and project members, hence the applicability of the strategies to these contexts took on a central role. Thus, what the district wanted and needed was deemed as more important than what benefited *Constructo*, which was clearly reflected in the discourse used. As expressed by one district manager: '[The discussion concerning the district's own priorities] is what is most important. We have to know what we want.... The business plan may be one issue. However, for us the discussion is perhaps more important. While top management rarely mentioned the words commitment, engagement or learning these terms and synonymous expressions often appeared in both the meetings and our interviews at the district level. According to one district manager, the failure of strategy implementation in the organization was often due to the lack of pedagogical skills among top management and their inability to 'sell' the strategy to the lower levels of the organization. 'They [Top management] use the wrong pedagogy when trying to sell [the strategies] by arguing from a top management perspective and not from a project perspective' (District manager).

At the district level, managers' attitudes toward strategy work were that it consisted of a constant negotiation in which the strategies had to be properly packaged in order to be effectively 'sold' to the rest of the district. This packaging meant that the strategies had to be perceived as being beneficial to the district first and foremost and then to the organization as a whole. One district manager explained his role in the following communication: 'My role is obviously to implement [the strategy], to break down and adapt it to our business. That is, to create commitment for [the strategy] based on our needs, the way we work and the people that work here.' As an example of the increased focus on commitment, one district started the process of working with the strategies through a two-hour brainstorming session where all the attendees were asked to write down the tasks they thought to be most important. This was done prior to the review of strategies advocated by top management. Based on the resulting list of tasks, decisions were made regarding the strategies to be prioritized. The district manager used this method in order to create commitment in the team: 'I want commitment to the strategies we prioritize in my district' (District manager).

This is in line with the statement of another district manager who believed that the most important outcome of the strategy work and development of the business plan was the dialog within the management groups: 'by working with the business plan during a whole week in the year, I believe that we incorporate a lot of ideas from the strategies in our daily routines' (District manager). The outcome of each district's strategy process was a district business plan with action points adapted to suit the local context of the district. Action points that concerned projects were to be entered into the project plan for detailed execution.

Project level

At the project level, the district's business plan was adapted for project activities using specific action points. We did, however, perceive that attitudes toward the project plan and its importance varied among the project managers. Although some claimed that in the past few years the focus on strategies and strategic issues had markedly increased, the impact of this increased interest on the projects was less discernable. Neither business plans nor project plans were given much attention in the day-to-day activities. One reason for this neglect, as expressed by one project manager, was that he could not find anything in the business plan that inspired him in his daily work. Another project manager considered the strategies and business plan as enforced by top management for control purposes. Thus, they gained concentrated attention for a short period, but were then forgotten for the rest of the year. The reason for this state of affairs, he said, could be found in the wide gap between top management's intended work process and how work was actually realized.

> [Constructo] works at two separate levels. The strategic level of our organization is never present at the work site and does not care how things actually work out there. They have their ideas and opinions on how things should be done, but it is out there that the work is done most effectively.
>
> (Project manager)

As we can see from these brief glimpses into the strategy work at three organizational levels, different approaches were adopted and were largely dependent upon the personal engagement, objectives and competence of the various managers. In this chain of strategy communication, different ways of translating and packaging the messages have taken place, leading to different ways of identifying with the strategies. In the following section, we provide a brief example of this 'travel of ideas'.

Strategies and organizational levels

As the strategies traveled through time and space, from one level to the next in the organization, their meanings were altered through translation. Some aspects may have been lost, while other aspects may have been gained. To exemplify this journey, we will follow one strategy as it travels, namely '[We shall] communicate actively with the outside world'. This strategy was defined in the strategy documentation as: 'External communication via for example the media constitutes an important channel for creating credibility and knowledge about our development work, our values and our collective competence' (Excerpt from strategy documentation). The rationale behind the strategy was explained by a member of the operational development team:

> The background for this strategy is that we want to measure our media image ... [The indicator] is the amount of positive mentions in the media in relation to our total media exposure.... We are simply not very good at projecting our image and doing so is important.
>
> (Member of the operational development team)

This strategy had already caused some tension at the regional level. While the regional manager's choice of words projected the importance of this strategy and pointed out the region's weaknesses in this area, he failed to translate this urgency to the district managers. The district managers did not view communication with the press as being part of their responsibility, especially since there was a special organizational function for this kind of task. The regional manager, however, did not agree with the interpretation of the 'outside world' as only including the press, but advocated taking part in debates or participating in research projects. The plausibility of these alternatives for the district managers, with their heavy schedules, could be debatable. The discussion ended in the decision to prioritize the strategy at the regional level and as an action point the district managers were to contact *Constructo*'s information department whenever they had something newsworthy to share.

Even though the district managers agreed to prioritize the strategy at the regional meeting, in their own meetings they continued to struggle with its meaning. Several district managers could not understand how they could operationalize the strategy. To them it was far from clear how the strategy would add value to their business. In fact, they were wary of conveying contradictory messages to the general public via the media and felt that communication with the media had not been part of their training. 'I am supposed to manage my business, be profitable and manage projects. I can, of course, take part in the external communication, but it is not my responsibility' (District manager). 'It is this indicator especially and this strategy that I feel I have no way of controlling. They both feel very distant from the activities in our district' (District manager). According to one district manager, the formulation of the strategy failed to convey what he thought to be its actual purpose: 'What we really want with the communication is to create a positive image of *Constructo* as a responsible company that wants to take part in the development of a sustainable society. But the strategy does not say that' (District manager). Even though the strategy was prioritized in most of the districts, little was done during the year to realize it. When failing to grasp the rationale and purpose of the strategy, as applied to their business, the district managers assumed a 'business as usual' attitude toward the strategy, which resulted in its being ignored or being paid lip service.

In this example, we see how the meaning of a strategy shifted as it was translated from one level to another in the organization. Although this strategy was deemed to be very important for the company as a whole, both the original formulation and the translation at the regional level failed to convey this urgency to the district and project levels. We have tried to show that the reason for this communicative mismatch does not only reside in the choice of words. The interlocutors' perspectives on the role and function of strategies as well as their loyalties – organization versus unit – also play important roles.

Discussion and conclusions

At top management level, strategies were considered a vital tool for organizational planning and control. The strategies were formulated by top management without involving other levels and were then presented to the managers, who were to disseminate them in the company. The strategies were in a sense regarded as orders that were to be acted upon by a capable workforce who shared top management's values and assumptions. Top management also established the strategic work procedure for the different levels. These characteristics correspond well with the definitions of the classicist approach to strategy as described by Whittington (1993). Moreover, this perspective correlates with the transmission model of communication and the abstract rhetoric used by managers in the studies of Johansson (2003), Fairhurst *et al.* (1997) and Müllern and Stein (1999).

As the strategies traveled down the chain of command at *Constructo*, we saw a marked shift in perspective on them. At the district level there was concern for the applicability of the strategies and their value for the local unit. The strategies had to fit the current local contexts if they were going to generate commitment and action. Likewise, more weight was given to how the strategies were formulated and interpreted. At the district management level, strategy implementation was viewed as a continuous negotiation process, both up the chain of command as well as down. The content and intent of the strategies had to be understood by those who were going to execute them.

At the district level, the process of developing a business plan served mostly as an opportunity for reflection on the local business for the coming year. Thus, the continuous negotiations to find an adequate fit between the needs and resources of the local unit and the organizational strategies reflected a *processual* approach to strategies as described by Whittington (1993). In the same way as the classicist approach, the *processual* approach influenced the ways in which the strategies were conveyed and interpreted.

The varying perspectives on strategies at the different levels in the organization seemed to be deeply rooted in the mindsets of the managers at those levels. The clash in perspectives renders understanding impossible, which in turn negatively affects the prospects for a successful implementation process. Formal planning, as viewed by classicists, is not however likely to disappear as a management tool in the foreseeable future. For example, the concept of analyzing and planning is very appealing on paper, gives authority to decisions and may function as a security net in a turbulent environment (Whittington, 1993). Since the *processual* perspective of the lower management levels is just as likely to prevail, there is a need to merge these two perspectives by combining resources from both.

The diametrically opposed approaches to strategies at *Constructo* created communication barriers, which resulted in the realized implementation process differing from the intended strategy process. Even if the lower organizational levels were not pursuing entirely different strategies than those advocated by top management, the fact that they approached strategies differently had practical implications for the *outcomes* as well as the *processes*. So, what implications might these heterogeneous approaches within one and the same organization have, for example, for the way strategies are interpreted and become embedded in an organization's activities and discursive practices? Also, what may be done to facilitate the strategy implementation process?

First, top management needs to be aware of the fact that an implementation process is not equal to a diffusion process (Rogers, 2003), which presupposes that the diffused idea, i.e. the strategy, does not change as it moves from one organizational level to another. Instead, strategy implementation could be seen as a translation process (Czarniawska and Joerges, 1996), where meaning changes as the strategy moves from sender to receiver. Different groups' or individuals' mental frameworks or mindsets, influence the way strategies are framed within a specific context as we have seen in this study (see also Weick, 1995); that is, how strategies are interpreted and understood is a context dependent process. This view of the implementation process can be compared to the systemic approach to strategy (Whittington, 1993). That is, the organizational members' behavior is rooted in their social networks and consequently the strategies reflect the social system in which they are enacted. The systemic approach does not, however, address the presence of different approaches within one and the same organization.

Without a common understanding of the strategies and how they are to be managed in an organization, there is a risk that tensions in both interpretations and commitment among the organization's members may prove destructive for strategic work. In order to achieve such an understanding, the different levels need to understand the underlying forces that influence interpretation and drive action. The top management levels above all need to acquire better knowledge of the affordances and constraints associated with the cognitive aspects of organizational communication in general and strategy communication in particular. While our study did not find evidence of any counteractions among the organization's managers, the way that the '[We shall] communicate actively with the outside world' strategy was approached illustrates that some managers distanced themselves from the strategy and responded with no action.

The different approaches to strategies at various organizational levels may not per se obstruct mutual understanding. It is rather the lack of awareness of the fact that other groups may have different ways of approaching strategy work that creates obstacles in understanding others' rationale for acting. For the effective implementation of strategies, different groups in an organization need first to understand the meanings embedded in the strategies and then accept that there are different ways of enacting them. Such an acceptance can only be achieved through interaction and communication. Our contribution has been to highlight some of the intangible underlying reasons for the mismatches.

This chapter has shown how different perspectives on the roles and functions of strategies in organizations give rise to different approaches to strategy work. In turn, these different approaches influence the ways in which strategies are interpreted and understood at the different levels. The ways in which strategies are understood thus influences the manner in which they are acted upon. Our intention has been to raise awareness of the complexity of strategic communication and encourage further research in this area.

11 Organizational change in the house-building sector

Abukar Warsame

Introduction

Two basic and related questions about industry in general is why production is organized in a particular way, e.g. what is produced in-house and what is outsourced to external firms? Unfortunately, this binary question is an oversimplification of a complex situation and cannot be used to cover the many possible alternative strategies. Lansley (1994) argues that a construction firm can be seen as a broker of opportunities for projects and as an intermediary acquiring resources for undertaking building projects. From a functional perspective, Tenah (1986) defines a construction firm as a group of people sharing specialized knowledge to design, estimate, bid, procure and obtain resources to complete a construction project. These functions extend beyond the boundary of a single legal entity and include an interwoven relationship with subcontractors, manufacturers, materials' producers and equipment suppliers. Thus, the interaction of these entities and how they transact their services and products shapes the organizational structure of a project and, ultimately, determines the governance structure of the specific firm (Shirazi *et al.*, 1996).

Winch (1989) argues that the primary object of construction management research should be the firm and the project should be seen as a temporary coalition of firms together with the client or owner. In line with Winch's argument, we will focus in this chapter on the organizational patterns of the firm – or group of firms – delivering the building project rather than the project itself. The objective of this chapter is to analyze various models of the organization of the firm within a particular sector of the industry – house-building – from the perspective of transaction cost theory. This approach to analyzing the house-building sector in terms of organizational patterns may enable us to understand better the bearers of risk and incentives, responsibility and control mechanisms and, consequently, it may shed light on the determinants of construction cost. The choice of this approach is motivated by the need to study organizational structure in terms of the nature of the relationships among participants in the construction process, rather than as an archetypical study of the kind that is often based on department groupings and management style of the organization. These groupings are often based on common tasks, products, geography and process (Grant, 2005).

The chapter presents some basic models of organizational structures in the construction industry. Four possible organizational forms will be presented. In the first model, the major players of the building process are integrated and market transactions are limited or non-existent. The other three models are basically an extension of this model and envisage a variety of organizational forms that could emerge when different firms in the process interact. Here, the prevailing structure is dictated by competition and market transactions. A later section explains a theoretical approach and the criteria for evaluating different organizational structures. Transaction cost theory in relation to various organizational patterns and the basis of three evaluation criteria will also be discussed. The evaluation of these structures is presented and followed by an appraisal of opportunities and challenges facing each type of organizational pattern in terms of flexibility and risk allocation, competitiveness and competence. Finally, the organizational patterns are summarized and future research directions identified.

Basic models of organizational structure

A house-building project is a multi-stage process where developers acquire land, consultants or in-house professionals carry out design and cost management, and (usually) contractors execute the construction. Others involved in the process are subcontractors, manufacturers and suppliers, although developers and contractors are generally regarded as the main actors in house-building; hence, all others work for them directly or indirectly. The client or owner of the land, working as a developer, might be building houses or apartments for sale or rent.

The first model represents an organization that has the ability to construct its own buildings with few or no resources brought in from outside. This simple arrangement is termed owner-developer-contractor (ODC), where these three actors constitute a single organizational unit. The firm has both the human resources and the physical assets needed to undertake building projects from land development to planning and design to construction, including all intermediate stages. Such an enterprise has a centralized, hierarchical structure and allocates its resources, products and services internally by administrative means rather than resorting to the market. A case that would fit this characterization is a municipal company (public sector organization) of a kind that could be found in Sweden in the 1960s, but which has since disappeared.

A typical structure that has also been found in different markets is where a separate firm carries out the construction work. There is competition between different contractors and the one selected builds to an order placed by the owner-developer (Figure 11.1). In this model, it is assumed

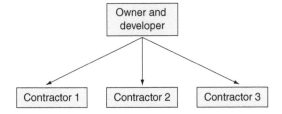

Figure 11.1 Model OD-C: owner-developer and a separate contractor.

that the contractor carries out all work using its own human resources. One possible scenario is that the construction part of a previously integrated firm (ODC) might have been divested and now competes with other firms for this and other work. The relationship between the owner-developer and the contractor can differ considerably depending on the specific contractual form.

In the next model – called OD-C-SC – the contractors in the above model have outsourced considerable parts of their work to subcontractors, who are also appointed after some kind of competitive tendering process. These subcontractors can work for any one of the main contractors.

Subcontracting firms can be active in almost any stage of the work, possibly starting off as consultants involved in various investigations and later in executing work on and off site. The model shown in Figure 11.2 has, for many years, been the dominating model in the Swedish house-building sector. In reality, there might be further levels of subcontracting where one subcontractor hires another to carry out part of the work for which the first firm has a contract.

Two more models can be observed. In Figure 11.3, the owner-developer hires the subcontractors – known as specialty contractors – directly instead

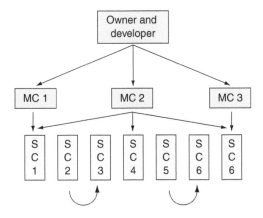

Figure 11.2 Model OD-C-SC (MC = Main Contractor; SC = Subcontractor).

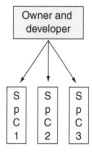

Figure 11.3 Model OD-SpCs (SpCs = Specialty Contractors).

of going through a (main) contractor. Figure 11.4 shows the owner-developer hiring a consultant to work on finding and coordinating the subcontractors. There is assumed to be a market for these consultants and competition between different consultancy firms. The next step is to look closer at the theoretical framework that will be used in the comparison of the models.

Theory and criteria for evaluation

Transaction cost approach

Transaction cost analysis can be undertaken on different levels. Robins (1987) states that the purpose can be either to explain the prevailing institutional structure of society at some point in history or the adoption of a specific organizational form in response to conditions faced by an individual firm. It is this second type of analysis that is reflected in this

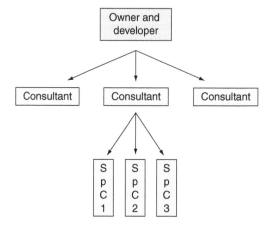

Figure 11.4 Model OD-Cons-SpCs (SpCs = Specialty Contractors; Cons = Consultant).

chapter with focus on three specific aspects: the flexibility of the organizational model in situations where the external market changes, the strength of competitive pressure on the activities carried out and, finally, the kind of competences the model demands from the client. The basic hypotheses are that organizational change can be seen as a reaction to problems in one or more of these three dimensions, and that the new model can handle these problems better and lead to a more efficient use of resources. New circumstances will, however, sooner or later lead to problems for this model too. In this chapter, examples from the house-building sector in Sweden are used to illustrate the theoretical arguments presented.

Lansley (1994) emphasizes the significance of the link between the firm and its environment and how the transaction cost approach is useful in explaining issues such as subcontracting, procurement method, and horizontal and vertical integration. Williamson's (1975) transaction cost theory, a development of the ideas in Coase (1937), is one of the most important tools that can be used to explain the practice of different contractual forms and procurement methods. The transaction cost approach explicitly regards efficiency as a fundamental element in determining the nature of organizations (Ouchi, 1980). In response to ever-changing business and economic conditions, construction firms adopt different kinds of organizational structure to influence procurement methods and better economize on the transaction costs of carrying out building projects. Morris (1972) points out that the effectiveness of the construction process lies in the management of the dynamic interrelationship between various organizations found on a building project.[1] Briscoe *et al.* (2004) noted that an organization's business environment and the adopted procurement route will affect the level of supply chain integration which, in turn, will affect future procurement decisions.

Criteria for evaluating organizational structure

Different environments, which generate different scales of uncertainty, require varying degrees of separation of organizational units and, likewise, different degrees of integration. Differentiation and integration of construction firms have some bearing on how risk is allocated and how firms respond to economic environment changes (i.e. their degree of flexibility). Nahapiet and Nahapiet (1985) contended that contracts represent different organizational arrangements for defining and coordinating the contribution of firms involved in a project. Reve and Levitt (1984) stated that each organizational pattern or governance structure corresponds to a particular type of contract ranging from classical contracting in market governance (contingent-claims contracts) to relational contracting in organizational governance (employment contracts). Walker (1996) refers to Lawrence and Lorsch's study (1967), which states that there is no one best way to organize, but rather the organization is a function of the nature of the task to be carried out and its environment. The level of separation and integration of the actors in the construction process can present many opportunities as well as challenges for each member of the construction project coalition. Grant (2005) claims, *inter alia*, that the lack of vertical integration in the construction industry partially reflects the need for flexibility in adjusting to cyclical patterns of demand and the different requirements of individual projects.

Child (1972) argues that environment, technology and size provide a powerful explanation of variation in organizational structure. He regards contextual factors as the most important determinant of structural patterns. The character of the business environment in which an organization operates – such as the nature of competition – has a measurable effect on organizational structure, job design and management (Brooks, 2006). It seems to be a difficult task to provide a unified theory or approach that fully explains the basis of organizational structures (Bridge and Tisdell, 2004). Thus, a combination of economic theory (transaction cost) and organizational theory (resource-based view), together with contingency theory, may provide some understanding of organizational structure in the house-building sector. As a method of general evaluation of different organizational structures, three rather broad criteria were chosen as they are in line with the transaction cost approach and seem to be relevant for the current issue: flexibility and risk allocation, competition and competence (see Table 11.1).

1   **Flexibility and risk allocation** refers to the degree to which an organization is able to respond to a changing economic environment and the ease with which it can utilize its resources efficiently. Uncertain factors that different parties in construction face can be categorized as risk (Ahmed *et al.*, 1999). How risks are allocated plays a major part in the

Table 11.1 Criteria for evaluating organizational structure and their definitions

Criteria	Definitions
Flexibility and risk allocation	Degree and ease by which major construction project parties handle uncertainties posed by changes in the economic environment and level of demand. It measures an organization's ability to adapt to changes in the economic environment.
Competition	Degree to which each organizational unit or subunit is put under competitive pressure.
Competence	The level of competence that an organizational structure requires and the opportunities for the unit and subunits to keep this competence continuously.

total cost of construction. Typical risks that contractors and owners often try to allocate or share by using various contractual conditions are delays, quality deficiencies, unit price increases and design changes. Construction firms are often confronted with uncertainties that arise from workload fluctuations due to the general business cycle of construction activity and the amount and size of contracts awarded. Workload dynamics may necessitate certain forms of organizational structure in order to handle not only risks stemming from the business cycle, but also successfully tendering for and managing projects with the optimal allocation of resources.

2   **Competition** – there may be competitive pressure on the parties in the construction firm. In the long run, if an organizational structure leads to a situation where more units of process have to compete with others for work then it will lead to higher efficiency. From an incentive perspective, Grant (2005) notes, for example, that vertical integration gives rise to what are termed low-powered incentives due to the internal supplier–client relationship that is governed by the vertically integrated organization instead of the market with its high-powered incentives.

3   **Competence** – different organizational patterns entail various degrees of competence in order to maintain any edge over an equal competitor. There could be several competence issues that could emerge from the formation of different organizational patterns. One of them is the particular set of competences required by firms in order to carry out their work in the project as efficiently as possible. In an efficient organizational structure, all firms will have the right competences and be in a situation where it is possible for them to keep individual competences up-to-date. It has been argued that competence problems on the developer's side played a major role in quality problems during the late 1990s. As construction activity had been very low in the early 1990s, developers did not have enough experience and know-how to cope with the challenges.

Evaluation of organizational structures

Some of the difficulties

A construction firm cannot gain sustained competitive advantage over another, since competitive pressures force firms to be more-or-less similar in efficiency (Ball *et al.*, 2000). If that is not the case, the result might be inefficiencies that induce the emergence of new organizational patterns. Integration of firms in the process increases the prospects for an individual firm to raise its profits. Barlow and King (1992) claim that the increased use of vertical integration is an alternative solution, enabling firms in Sweden to manipulate production costs. Integration of developer and contractor with specialty contractors might increase the competence of the integrated organization, but it may also limit flexibility in adapting to economic changes. In contrast, a separate developer, contractor and specialty contractor organizational form may allow firms to act competitively in the prevailing economic environment, leading to a better risk allocation; however, the required competence of each firm may have to increase in order to engage in the most efficient way in the contracting process. Reliance upon consultants or the use of other forms of contracting could arise in the absence of an essential competence.

Eccles (1981a) notes the fundamental questions that vertically integrated firms face, which include the extent to which a firm is responsible for providing all the required input resources to produce output and the basis for organizing work and managing relationships with other firms should it have chosen to obtain inputs from them instead. Smaller organizations are characterized by centralization of power for formulation of strategy and adaptability in order to respond to economic changes (Shirazi *et al.*, 1996). However, non-integrating, smaller organizations may suffer from limited resources in the face of large projects or during periods of high construction activity which tend to favor large firms. By the same token, large firms with abundant resources may face the reality of economic downturns where construction activity and demand are low and many of them struggle to utilize their resources efficiently.

Since most construction work is obtained on the basis of competitive tendering (Gonzalez-Diaz *et al.*, 2000), construction firms often face the difficult task of choosing those projects where they are likely to be rewarded. Firms tender at a price that would not only allow them to win the contract but also to provide them with a margin of profit. Similarly, construction firms – big or small – have different capacities and capabilities for undertaking one or a few contracts simultaneously and that itself creates a source of uncertainty. There are also occasions when a general contractor cannot retain the services of a large number of specialists because of uncertainty concerning the requirement for labor as dictated by time, location and specialty (Eccles, 1981a). The bulkiness of construction material affects transportation costs and can result in a regionalized market structure (Lowe, 1987) that reduces the flexibility for transferring materials to where they are most needed. One of the major advantages of subcontracting is in reducing these uncertainties and passing much of the risk to the subcontractor, given that the subcontractor has more opportunities to handle these risks than the general contractor.

The potential organizational change that any of these criteria – *flexibility and risk allocation, competitiveness* and *competence* – could bring about is not the same. The need for higher flexibility and better risk allocation in an unstable economy and volatile construction activity is much more serious than addressing a higher competence requirement or competitive pressure. These criteria can be formulated as hypotheses concerning changes in organizational structure. Such changes are likely:

- when there are other organizational forms that can manage risk better;
- when there are other organizational forms that are more efficient because more units are put under competitive pressure; and
- when the current organizational form needs competences that are difficult to maintain in the organization compared to the position for other organizational forms.

From the ODC to OD-C model

The owner-developer-contractor model faces many challenges as well as opportunities, ranging from increased risk exposure and bureaucratic costs to an improved level of competitiveness in terms of capacity and less reliance on other firms to provide the desired inputs. Uncertainties in the development and construction markets are present simultaneously in the ODC organizational model in addition to the risks that arise when the organization is not engaged in subcontracting. A more unstable economy also causes under-utilization of resources amassed by this type of firm, because transfer of labor and materials from a low to high demand region is uneconomical.

The ODC organizational type could improve the competitive position of the firm and the competence of its entities. Organizations grow by acquiring or merging with other firms in order to benefit from economies of scale or scope. Thus, one can assume that a vertically integrated organization with capital and manpower muscle has improved its competitive advantage. Even so, small firms and specialty subcontractors have advantages over large firms for many small jobs and repair works (Foster, 1964). Many of the competitive advantages of large firms – wealth, superior technical knowledge and the scope of standard procedures – may prove ineffective for small projects and a variety of other situations encountered (Foster, 1964). Thus, an organization must be large enough to compete and at the same time small enough to specialize in certain types of construction work.

The main problem for the ODC organization from a competition perspective is that of control problems, meaning that various parts of the organization might not work efficiently because they are sheltered from competition. The differentiation between owner-developer and contractor presents the developer with an opportunity to deal with other contracts and to practice market-deriving procurement. Nevertheless, the separation compels the developer to acquire the necessary competence to undertake the 'nuts and bolts' of the transaction process efficiently – from the design and specification stage to tendering and procurement of the final product/asset. The OD-C model might replace the ODC model for two reasons.

1 Independent contractors can handle market fluctuations better than the integrated ODC firm.
2 The contractor is put under more competitive pressure as the developer can choose between different contractors. There could be a reduction in administrative costs as result of splitting the work, as well as efficiency arising from transacting at arms-length with high-powered market incentives.

From the OD-C to OD-C-SC model

The opportunity to transform fixed costs into variable costs through subcontracting practices is absent in the OD-C model. Thus, this form of organization could still be associated with a lack of flexibility and inability to allocate risk. The OD-C-SC tries to rectify these weaknesses. A reduction in overhead and construction costs stemming from lack of local market knowledge and the need for supervision are some of the benefits that contractors gain from subcontracting.

By subcontracting to specialist firms, contractors can avoid committing themselves to significant investment in terms of labor and other assets. Thus, subcontracting allows (main) contractors to secure numerical and functional flexibility (Velzen, 2005). An important, but rather neglected, issue is how subcontractors can handle market risk. There are several possible answers. Subcontractors can of course work for different developers and they might also be able to work in large regions. They can also cooperate informally with other subcontractors if they have excess supply or excess demand for specialists. They might also work on maintenance projects and directly for owner-occupiers in the housing market.

The OD-C-SC organizational model seems to provide greater flexibility and risk allocation for firms by allowing for both market contraction and separation of activity between development and construction. There still might be a problem from lack of competition in the marketplace compared with the OD-C model. If that is the case, developers face competitive pressures to:

- integrate with non-vertically integrated firms or establish a contracting section within their organization;
- protect against opportunistic market driven behavior with painstaking use of contracts; or
- gravitate toward direct contracting with specialty contractors.

The competence required of the developer in the first scenario above is not so demanding unless it includes procuring projects externally. The other two scenarios do, however, demand higher competence in order to tender, manage, monitor and evaluate the performance of contractors.

From the OD-C-SC to OD-SpCs or OD-Cons-SCs model

Developers have no direct contract with subcontractors in the previous forms of governance. Although they are not the real beneficiaries of subcontracting, they remain exposed to the risk of subcontractors underperforming. When regular developers act as owners-developers of projects, they have the opportunity to work repeatedly with specialty contractors and build long-term relationships. Specialty contractors are not only an integral part in new projects offered by the developer, but also have the opportunity to carry out repairs and renovation works on older projects owned by the same developer. This can mean a reduction in transaction costs (search, administration, contracting and monitoring costs) for both parties. Learning by doing is an old philosophy from which the owner-developer can benefit. Developers may accumulate the skills and experience needed to carry out future projects without the employment of agents.

The situation is somewhat different when the developer is not regularly engaged in developing projects. On the one hand, specialty contractors may not be able to establish a good working relationship with the developer, and vice versa, and thus trust and harmony may be in short supply. The developer might also not have the skills to find the right specialty contractors and then coordinate their work. On the other hand, the developer might not want to use the OD-C-SC structure because of the lack of competition between main contractors. In this situation, the hiring of a project manager or consultant firm that could provide the necessary skills is one possible alternative for procuring the services and products of subcontractors. In a competitive market, a good working relationship would obviously serve the best interests of both the infrequent developer and the consultant. The infrequent developer needs to utilize the expertise and resources of consultants – something that regular developers should already be enjoying.

In this organizational model, the owner-developer's competence and the increased use of subcontracting in order to allocate risk might not be enough to compete against larger, vertically integrated firms carrying out major projects. Ball *et al.* (2000) have stated that banks and insurance companies may not be able to monitor contractor performance and rely on size as a proxy for competency and solvency. Thus, larger firms may have an advantage over smaller firms when it comes to the financing of projects. The competition could be even more acute in major cities or a growing region where a high concentration of large contractors is likely to be present. Often these owner-developer firms operate at the regional level and in medium-sized cities where the developer and few specialty contractors could develop a dyadic relationship rather than a market-driven relationship.

Other developments

There are probably many reasons why different firms choose to integrate or differentiate. Turner (1997) has pointed out that depressed economic situations and reduction of general housing subsidies in the private rented sector forces municipal firms to be more efficient and more businesslike in order to be able to handle a harsher economic situation. In Sweden, the number of firms that are active both as developers and contractors is very small, although there are a few more developers or contractors of a medium size.

It is not only developers who are forced or motivated to integrate; contractors could also initiate integration. A contractor that wants to enter the rented housing market or build its own projects in order to maintain efficient use of its resources may integrate with a developer or create a separate division that fulfills these objectives. A survey carried out by Warsame (2006) found that some developers in small regions of Sweden have raised concerns about large, vertically integrated contractors entering the rented housing market which had already experienced a lack of competition among large contractors. One of the motivations for large, vertically integrated contractors engaging in the rented housing market could be their anticipation of subsidy policy changes.

Conclusions

A better understanding of the various organizational structures as well as the economic and market forces that determine their efficiency could not only help to expose cost differences across regions, but it could also assist in predicting the kind of organizational structure that might emerge in the future. Will the dominant position of large contractors be broken by the use of consultants and specialty contractors? Will large contractors then respond by being even more active as developers?

The use of transaction cost theory as a tool for exploring different organizational structures from an efficiency perspective makes it easier to predict how firms engaged in projects respond to economic and business challenges that are vital for their survival. It may explain why particular organizational structures have dominated at some point. It also helps to reveal forces (competitive pressure, higher level of required competence, greater flexibility etc.) that make it necessary to consider another form of organization.

Four main organizational models were envisioned where firms in the construction process are allowed to integrate or separate in response to uncertainty and changing business cycles. The owner-developer-contractor model faces many challenges as well as opportunities ranging from increased risk exposure and administrative costs to an improved level of competitiveness in terms of capacity and less reliance on other firms to provide the desired inputs. A more unstable economy also causes under-utilization of the resources amassed by this type of firm where the transferability of labor and materials from a low to high demand region is uneconomical.

Separate owner-developer and contractor structures with permitted subcontracting practices might provide risk reduction and greater flexibility. Risk is allocated to those who can handle it best. Firms are able to enter market arrangements that would allow them to minimize their exposure to market volatility. Dealing directly with specialty contractors or engaging in subcontracting enables firms to reduce overheads and construction costs stemming from uncertainties of workload and lack of local knowledge. The use of consultants could rise where developers infrequently undertake projects and thus lack the resources and skills necessary to carry out projects.

The organizational patterns presented in this chapter as well as the presumed differentiating factors (flexibility and risk allocation, competition and competence) that initiate changes in organizational structure have not been empirically tested. The author therefore intends to carry out a survey in Sweden and countries with similar housing markets and policies. The study will explore the effects of exogenous environment factors, namely competition and uncertainty, and endogenous factors such as the developer's required competence in relation to current and past organizational structures.

Note

1 Cited in Shirazi *et al.* (1996).

12 Trust production in construction

A multilevel approach

Anna Kadefors and Albertus Laan

Introduction

In many countries, the construction industry is criticized for being adversarial, inefficient and conservative (Latham, 1994; Egan, 1998; Ericsson *et al.*, 2002; PSIB, 2003). Low trust is often seen as a major cause of inefficiency problems, and more collaborative procurement routes are advocated. In literature on trust, it is generally stated that an appropriate level of trust has important benefits for inter-organizational cooperation and relational performance. Research on inter-organizational trust appears, however, to be full of paradoxes (Nooteboom, 2002). For instance, trust can be based on contracts and control, but can also rely on affection and norms of reciprocity and fairness, in which case formalized control sometimes may hamper trust. It may enable openness and flexibility, but can also be so strong that it limits the variety of business relations needed for learning and innovation (Nooteboom, 2002). Furthermore, trust does not only arise in direct interaction in specific exchange relationships, but is also influenced by more general contextual characteristics. The contracting environment, involving legal and educational institutions as well as ethical norms and other cultural aspects, interacts with formal contracts and the behavior of individuals in shaping trust development in a specific relationship.

The purpose of this chapter is to identify major bases of trust production in the construction industry. We assume a multilevel approach, focusing on the interaction between institutional and relationship-specific foundations of trust. A starting point for this chapter is that an industry or organization is not only structured to manage threats of opportunism, but also to coordinate the actions of different parts of the organization efficiently. In construction, where temporary organizations composed of a multitude of firms jointly produce unique and complex buildings, the need for information processing for coordination purposes is very high. Thus, in our analysis we assume that two parallel systems shape both industry level institutions and project level relationships: one system for information processing, coordination and knowledge management, and one system for mitigating opportunism and creating trust (cf. Madhok, 2006). The systems are partly interrelated, so that goals of information processing may sometimes produce structures and actions that interfere – positively or negatively – with goals of trust and relationship building, and vice versa.

In this chapter, we identify key aspects of these two systems in construction and discuss how they interact on different levels. First, we identify relevant forms of trust and processes of trust production, based on trust literature. We then briefly introduce central concepts in theories of organizational structure and coordination. Subsequently, we describe key characteristics of the construction industry and discuss their relationship with information processing and trust production.

Literature review

Forms of trust and processes of trust production

The growing amount of research into trust undertaken over recent decades has clearly established that there are several forms of trust, linked with different processes of trust production (Rousseau *et al.*, 1998; Bachmann *et al.*, 2001; McEvily *et al.*, 2003). Economists (e.g. Williamson, 1985a, 1993) tend to view trust as calculative, based on conscious evaluations of the trustee's objective self-interest and competences, while social scientists (e.g. Granovetter, 1985) emphasize the socially embedded character of trust, shaped by emotions arising from interpersonal relationships. Rousseau *et al.* (1998) suggests that trust can be considered as a meso concept, integrating micro psychological processes with macro level institutional arrangements.

Starting on the society level, institutions are formal and informal social structures that promote trust by providing stability and predictability. As such, educational systems, legal frameworks and various kinds of established practices can act as broad supports for trust within exchange relationships (Sitkin, 1995; Lane and Bachmann, 1997; Lindenberg, 2000). The role of a legal system is not so much to punish breaches of contract as to facilitate cooperation by creating confidence in contracting as an organizing principle (Lane and Bachmann, 1997). In a specific industrial context, bilaterally agreed standard contracts may have a similar role, also creating confidence in the fairness of contractual terms (Vlaar *et al.*, 2007; Arrighetti *et al.*, 1997). Arrighetti *et al.* further emphasize the importance of the contractual environment, referring to a broader context of laws, customs and assumptions that includes market structure as well as prevailing notions of what constitutes ethical behavior in business relations. Also vital for trust in exchange relationships are educational institutes, professional associations and normalization bodies, which provide certification of individuals and firms, pertaining to their competence, systems and sometimes also ethical conduct. Informal institutionalized beliefs, such as the perceptions that people in general hold regarding, for example, the trustworthiness and competences of various occupational groups or appropriateness of different behavior, further contribute to structure interaction and influence trust.

In moving to the level of a specific inter-organizational relationship, a certain degree of institutions-based trust is required for a relationship to be established at all. Further, when the parties first meet and no relational experience has taken place, initial levels of trust will be partly based on institutionalized expectations. Another basis of trust in this early stage is an assessment of the other's interests and opportunities for actions. Such calculus-based (Rousseau *et al.*, 1998) or rational trust (see, for example, Nooteboom, 2002) involves considerations of the other party's dependency and prospects for short-term gains and future exchange, as well as of how contractual arrangements and means of control influence opportunities for opportunistic strategies. Thus, as noted by Lane and Bachmann (1997), these early risk assessments are strongly based on the prevailing institutional order. Both institutions-based trust and calculus-based trust are considered to be weak or 'thin' (Nooteboom, 2002) forms of trust, the sources being impersonal, not involving direct personal interaction and experience. As will be further discussed, some authors do not see calculus-based trust as a form of trust at all, but rather as calculated risk-taking (Lindenberg, 2000; Nooteboom, 2002).

When an inter-organizational relationship is started and develops, the weaker forms of trust can be extended with, or replaced by, a form of trust that (although with slightly different meanings) has been labeled relational trust (Rousseau *et al.*, 1998), process-based trust (Zucker, 1986) or affect-based trust (McAllister, 1995). In a relationship, feelings of personal attachment and tacit mutual understanding will arise and influence actions taken (Ring and Van de Ven, 1994), and this type of trust is also described and seen as a strong or 'thick' (Nooteboom, 2002). Foundations of such relational trust are perceptions of benevolence, empathy, openness, loyalty and dedication (Mayer *et al.*, 1995; Klein Woolthuis *et al.*, 2005). Accordingly, Nooteboom (2002: 48) defines trust as 'an expectation that things or people will not fail us, or the neglect or lack of awareness of the possibility of failure, even if there are perceived opportunities and incentives for it'. Thus, stronger forms of trust require that trustworthiness goes beyond what is prescribed in the formal contract.

It should also be mentioned that strong trust is not always beneficial. There is both a risk of betrayal and a risk that relationship loyalties interfere too much with other business goals (Lindenberg, 2000). Many inter-firm relations comprise a number of ties between several individuals in each organization and the levels of trust may vary significantly between them, depending on the frequency and type of interaction. Relationships that involve little contact will hardly bring about more robust forms of trust. Development of trust may therefore follow patterns of interaction and communication which, apart from needs for collaboration, are influenced by aspects such as geographic proximity, personal liking and individuals' time pressure.

It should also be noted that rules of reciprocity are central to trust development (Kramer, 1999). As shown, trust tends to be reciprocated by collaboration, while perceived distrust is likely to induce opportunism – and so produce the behavior it is intended to prevent – such that trust and distrust then become self-fulfilling prophecies. Trust may therefore be seen also a control mechanism influencing the behavior of an exchange partner in a more collaborative direction.

Trust, contract and control

In general, the relationship between trust and control is ambiguous and complex (Bradach and Eccles, 1989; Poppo and Zenger, 2002; Klein Woolthuis *et al.*, 2005; Long and Sitkin, 2006). As mentioned above, economists focus on calculus-based trust and see contractual safeguards and related formal control activities as conditional to trust (Williamson, 1985a, 1993); whereas, it is a central theme in many definitions of strong forms of trust (e.g. Nooteboom, 2002: 48) that there has to be evidence that a person (or organization) chooses to collaborate despite incentives to pursue self-interest at the trustor's expense (see also Malhotra and Murninghan, 2002). Rules of reciprocity also imply that the development of trust is hampered if control is perceived as a sign of distrust. Thus, research on the role of contracts in inter-firm relationships has tended to see trust and control as substitutes. In this view, trust reduces the need for contracts to the extent that contracts and other types of formal control are sometimes perceived as harmful to trust (Macaulay, 1963).

Empirical research has shown that contracts and other aspects of formal control may, in some cases, strengthen trust and that trust and control can thus be complementary constructs (Zaheer and Venkatraman, 1995; Poppo and Zenger, 2002). One reason is that contracts can act to focus partners' attention and increase their knowledge of the agreement, thereby promoting a shared understanding that both coordinates action and supports trust (Vlaar *et al.*, 2007). So, for an effective relationship performance, partner firms need to manage emerging risks and uncertainty adequately by understanding the conjoint roles of trust and control (Das and Teng, 2001).

Systems for information processing

Earlier in the chapter, we stated that systems for trust production and coordination are strongly interrelated. Systems and activities can be independent, but may also be either complementary or supplementary, much in the same way as the relationship between trust and control described above. Interdependencies can occur within organizations, within inter-organizational relationships and on the institutional level. One example is the dual role of contracts as devices to reduce risks of opportunism by defining obligations, monitoring schemes and punishments, and as key tools for communication. Similarly, a specific type of meeting can be organized to facilitate coordination, but the group of people involved will also start to develop relationships. This implies that to understand the system of trust production in an organization, we also need to understand communication processes.

For help in identifying organizational activities and structures related to information processing, we may use the conceptual framework of contingency theory (Thompson, 1967; Galbraith, 1973), where the need for communication and coordination is seen as the main determinant of organizational structures. Planning, hierarchy, mutual adjustment and standardization are important coordinating mechanisms, and their applicability is related to the type of interdependence between activities. Standardization may pertain to work processes, output, and skills and knowledge (Mintzberg, 1983). The latter form is important in professional work, where authority is delegated to individuals on the basis of their expertise, skills and often also a value orientation acquired by socialization within a professional community. Standard contracts serve the purpose of reducing the need to develop bespoke contracts and thereby economize on information processing. As noted by Vlaar *et al.* (2007), standard contracts may also entail less joint sense-making and thus reduce mutual understanding about the agreement.

In the following sections, we identify key bases of trust (and distrust) in construction and discuss how these relate to systems of information processing. The starting point of the discussion is that the organization of the construction industry is fundamentally shaped by the nature of its outputs.

Project-based organizing in construction

Buildings (and civil engineering works) are distinct from most other products in that they are technical monopolies, meaning that once a building occupies a piece of land it cannot be replaced without huge cost. Obviously, it is not possible simply to return a faulty building to the producer/ contractor, and important defects may be hidden and/or irreplaceable once the systems are in place. This irreversibility means that the client is highly vulnerable both to design deviations and quality defects. This is an important reason why clients traditionally employ strong organizations to control contractors.

Despite efforts to increase pre-fabrication, all buildings are inevitably to some extent prototypes. The same holds for construction project organizations, which are temporary multi-organizations (Cherns and Bryant, 1984), composed of numerous specialized units. Firms in the construction industry are generally tied to a specific system (e.g. structure, cladding, heating and catering) or, especially in the case of consultants, to a functional competence (e.g. aesthetics, management, fire safety, geotechnical and acoustical engineering). Organizing by trades in this manner has a long history (Winch, 2000) and today, when the complexity of buildings increases, project organizations may also become more complex. Thus, the immobility and complexity of buildings are two factors that both have important organizational implications. Immobility leads to site-specific, unique products and organizations, in turn resulting in high levels of uncertainty and high demand for inter-organizational communication.

The construction process is generally divided into distinct stages, and the division of labor varies according to the procurement route chosen. In a design-bid-build contract the client is responsible for the design, while in a design-build contract the contractor has this responsibility. In both cases, the output of the design phases takes the form of documents (brief, drawings and specifications), and the price of changes and additions after the construction contract is signed is set by negotiation and not by competitive tendering. Thus, changes are often costly to the client. Below, we discuss how these contractual rules interact with standardization of roles to influence trust building.

Discussion

Trust, contract and control

Transaction cost analysis based on transaction frequency, asset specificity and uncertainty demonstrates why costs for setting up and managing project-based contracts are high in construction (Winch, 1989). The industry is often mentioned as an example of a context with a hybrid governance structure, either because of long-term relationships between actors (Eccles, 1981b; Bradach and Eccles, 1989) or because contracts incorporate important hierarchical aspects (Stinchcombe, 1985; Bradach and Eccles, 1989). Such hierarchical elements are client control/inspection of site activities and, perhaps most important for trust, the exclusive right of a contractor to carry out additional work, due to changes and contractual omissions, on cost-reimbursable terms. The monopoly position of the contractor in pricing changes creates room for opportunistic strategies, especially in combination with procurement auctions with price as the main choice criterion. A contractor may therefore put in a low bid and expect to compensate it with income from claims for extra work (Rooke *et al.*, 2004). Another option is to shirk on quality, given the possibilities to hide work and to substitute high quality materials with cheaper, lower quality materials.

This issue has implications for both calculus-based trust and relational trust. First, because of the opportunities that the contractor has to take advantage of the situation, a 'rational' client often adopts a defense strategy, based on monitoring and close analysis of all claims (although large and regular clients may count on reputation effects moderating opportunism). As for contractors, there are rational reasons for their questioning the competence of the design team but also of the client, who has collaborated in producing the contract specifications. Thus, for both sides a certain level of distrust and opportunism can be defended on calculus-based grounds.

As stated in the earlier section on trust, contract and control, intuitive psychological rules of reciprocity imply a risk where distrust initiates a vicious circle and becomes a self-fulfilling prophecy. A contractor that is closely monitored by a suspicious client may reciprocate through an opportunistic claims strategy. Furthermore, a suspicious attitude will make the client more inclined to interpret all contractor suggestions as motivated by self-interest rather than by a concern for the project that would otherwise signal trustworthiness. Calculus-based defense strategies will easily produce behavior that is not compatible with goodwill, benevolence and loyalty that characterize developments of relational trust. This is not so much of a problem in relatively standard contexts, where specifications are simple, changes few and norms of quality undisputed.

When change negotiations are few and cause little conflict, traditional contracts are functional. Both trust and quality/efficiency ambitions then tend to be institutionalized on a moderate level, not requiring too much contract management. A design-build approach may further economize on transaction costs, since the number of negotiations is lower when specifications pertain to systems and functions while detailed design is a contractor responsibility. When uncertainty and uniqueness increase, however, traditional contracting becomes more risky as interaction is dominated by negative issues. At a certain point, a bad relational climate may be so costly that a different contracting philosophy is needed. This often implies that the client assumes more risk, and trust problems then affect the contractor's motivation to keep costs down. In such cases, control will be based not so much on mitigating opportunism by complete contracts and inspection but on informal, value-based management strategies.

Trust and information processing

In an industry where firms regularly form temporary constellations, it is obvious that industry-level standardization is more powerful than firm-level standardization when it comes to reducing the need for coordination. Accordingly, we find a variety of industry level standards in construction, pertaining to organizational aspects as well as to physical artifacts. For building components, national and international normalization bodies provide important output standardization. On the organizational side, there are standard contract agreements that regulate responsibilities as well as handling contingencies and specifying requirements. Following Vlaar *et al.* (2007) and Arrighetti *et al.* (1997), standard contracts may be conducive to trust in that they increase the confidence in contracting as an organizing principle and a belief that the contractual terms are fair. This is a likely effect of standard contracts in construction.

Strong expectations relating to knowledge and attitudes are tied to the various actors in the field, often based on a combination of cultural and formal characteristics. Consultants especially, but to some extent also craftsmen and contractors, are trusted because of their specialist knowledge and professional/crafts ethics (Stinchcombe, 1959; Bowen *et al.*, 2007). Both individuals and firms may be subject to certification and accreditation, performed by independent institutions. Thus, since standardized processes and roles provide both predictability and legitimacy, the same system that economizes on transaction costs also functions as an important source of institutions-based trust. Yet, when it comes to trust in and between different categories of participants, we should perhaps speak instead of institutionalized levels of trust. There is strong prejudice between some categories of participants, which in many countries is most pronounced in the case of architect–contractor relationships (Phua and Rowlinson, 2003; Loosemore and Tan, 2000). Institutionalized roles therefore provide a standardized level of weak trust, which is sometimes more adequately described as distrust.

This standardization of roles is a fundamental organizing principle in the construction industry. Although each project organization is unique in terms of the competences and specific companies involved, there is a limited set of organizational models (procurement routes) and only some competence combinations available in the market. Companies are designed to fit into specific slots in project organizations, and individuals and firms perform similar tasks in all projects to a large extent. An important function of this industry-level standardization of roles, knowledge and responsibilities is to reduce the need for communication and negotiation between parties. Although construction firms are decentralized, the freedom of project organizations is strongly constrained by contextual factors (Bresnen *et al.*, 2004).

The standardization of roles also has consequences for relational trust. In the traditional procurement route (and, to a lesser extent, in design-build options), the project is divided into distinct phases where output takes the form of documents (brief, drawings and specifications). As mentioned above, changes and additions after the contract is signed are expensive. Consequently, project control is strongly directed toward minimizing changes, aiming at producing documents that are as flawless as possible. This has a side effect in that participants in different project phases have little contact, and that any interaction which does occur tends to be conflict-oriented.

Standardization also reduces the need for communication within phases. When a specialist consultant or contractor is brought into a project, there is seldom an organized induction or team-building session; the new team member is expected to know what to do and to perform immediately.

Although trust does not always arise in interpersonal relationships, those without any interaction and communication will hardly bring about more robust forms of trust. That a construction project is sub-divided into phases and standardized work packages also impacts on the opportunities that different participants have to build relational trust.

A related aspect is how geographical movements in space affect spontaneous trust building. Co-location of teams often brings about strong interpersonal trust and efficient work practice (Scarbrough *et al.*, 2004), but this is only possible in larger projects. Instead, most design interaction takes place in scheduled meetings and there is little room for spontaneous socializing between individuals from different firms. This contrasts strongly with the construction phase, during which the construction site is the principal physical hub and meeting location. Here, the building contractor, as a main contractor and coordinator, is the key actor who has opportunities to build relations with other contractors as well as the client representative and other site visitors. Site-level trust may moderate distrust on higher organizational levels and is often considered indispensable to efficient problem solving. The result may also be compromised where, although not necessarily intentionally, the interests of consultants, users or even one's own company are overlooked. Thus, patterns of interaction contribute to the development of both loyalties and distrust.

Conclusions

A starting point for our discussion was that the products of construction – buildings – are the roots of the project-based organization in the industry. In order to handle vast amounts of uncertainty and reduce transaction costs in this context, extensive formal and informal industry-level standardization has been developed. Yet, to economize on human interaction by communication and mutual adjustment means an increased reliance on weak forms of trust, while the development of stronger, relational trust is hampered. While standard contracts might reduce the need for contract parties to communicate (Vlaar *et al.*, 2007), standardization of roles is perhaps a more important cause of absent joint sense-making across organizational boundaries in construction projects. As a consequence, relational management is underdeveloped compared to opportunism management. Left unmanaged, building of relational trust is limited to spontaneous socializing in the interaction that is brought about by the production system, mainly taking place on the construction site. There is reason to be more attentive to how such spontaneous, emergent trust-building influences project outcomes.

As for the relationship between trust and contract, the latter may seem to be responsible for much of the distrust in construction. A closer look suggests that it is not contracts per se that are problematic, but rather there exist uncertainties that are hard to solve through contracts alone. Clauses are introduced to handle change management, but changes remain important sources of conflict. Still, it is possible that the tensions produced by uncertainty could be better handled if relationships were characterized by stronger relational trust. This requires an approach where relationship and knowledge management assumes prominence over opportunism control. Furthermore, to improve trust and collaboration, project management needs to approach both formal and informal communication not only as means to coordinate action, but also as bases for strategic relationship building. This implies that more resources have to be invested in expensive project-level communication, partly related to unlearning of institutionalized behaviors developed in a traditional weak trust context. Prevailing institutionalized coordination mechanisms seem increasingly outdated when requirements for flexibility rise and many projects are too complex to handle with a weak trust approach. We should ask what kind of industry-level standardization would serve the needs of efficient coordination, yet be better adapted to the goals of flexibility and explorative learning?

13 Value-based award mechanisms

Marco Dreschler, Hennes de Ridder and Reza Beheshti

Introduction

The exclusive and prolonged use of a lowest price-based procurement strategy by public authorities can have a severe, negative impact on the construction industry as a whole. Despite this tendency, public authorities continue using this method, because they believe it is the only way to justify the spending of taxpayers' money. In the construction industry in the Netherlands – and in other countries too – this results in problems such as low client satisfaction, low profits, lack of innovation, high transaction costs, troubled client–supplier relations and even corrupt behavior, as concluded in previous studies by the authors (Dreschler and de Ridder, 2006; Dorée and de Ridder, 2003; de Ridder *et al.*, 2002).

A number of problems in the Dutch construction industry can be attributed to the dominance of the lowest price contract award mechanism in public procurement. In this kind of procurement, companies are forced to customize beyond the economic optimum, causing the industry to underperform. As a result of the lowest price-based procurement regime, contractors and, indeed, the entire supply chain have to comply with a detailed list of demands, setting up a new production system each time. In this regard, standardized components or company-specific solutions can only be applied with great difficulty, which means the supply chain cannot organize itself in the most economical way. Furthermore, the traditional organization model – an architect or engineering firm producing a design for a client (Roelofs and Reinderink, 2005) – which is associated with the lowest price selection mechanism results in designs lacking constructability, because design and construction knowledge and responsibilities are separated. In addition, most clients are not able to describe their requirements realistically or in detail, because they lack thorough knowledge of the market. This feature becomes a problem with the lowest price-based selection regime and, ultimately, results in high preparation costs and unrealistic expectations.

Problems associated with lowest price-based procurement are unrealistic customization demands, lack of constructability in designs and the need to 'predict the unpredictable'. Value-based procurement can solve these problems. Using this approach, it is no longer necessary to give a detailed description of the desired solution and so suppliers must assume more responsibility for the design. The approach also allows suppliers to apply company-specific solutions or components, which makes designs more constructible. Furthermore, the client no longer has to specify needs in great detail in advance. Hence, value-based procurement can bring about improvement from which clients, as well as suppliers, can benefit.

Several factors threaten the adoption of value-based procurement. The main reason is that public clients lack confidence in value-based procurement. They often hesitate to formulate award criteria, because even small details of these criteria have potentially large consequences. When the public client is faced with the uncertainty of legal consequences, in addition to a general lack of experience with the process, the attractiveness of the lowest price-based selection strategy becomes obvious. The objective of the study presented in this chapter is to increase knowledge of value-based procurement. As explained in a later section, this will be accomplished by a study of the implementation of the EMAT (economically most advantageous tender) award mechanism.

Value-based procurement

EU regulation and definitions

According to procedures[1] for the award of public works contracts, public supply contracts and public service contracts, public clients have two possibilities. Without prejudice to national laws, regulations or administrative provisions concerning the remuneration of certain services, the criteria on which the contracting authorities can base the award of public contracts are either:

(a) when the award is made with respect to the tender that is economically the most advantageous from the point of view of the contracting authority (based on various criteria linked to the subject matter of the public contract in question, for example, quality, price, technical merit, aesthetic and functional characteristics, environmental characteristics, running costs, cost-effectiveness, after-sales service and technical assistance, delivery date and delivery period or period of completion); or
(b) the lowest price only.

The word 'criteria' as used in the procedures has two different meanings. In the introduction 'criteria' is used in the sense of award mechanism. In point (a) 'criteria' is used in the sense of 'a property on which a performance evaluation is based'. This chapter uses the term 'award criterion' for the second meaning. Public clients usually use the award mechanism in the award stage where the final bids of suppliers are ordered and the best bid is selected (Figure 13.1). The award phase is part of an encompassing procurement strategy and usually comes after the selection phase – see OGC[2] example of a description for the entire procurement process.

Figure 13.2 portrays the lowest price award mechanism, while Figure 13.3 portrays the EMAT award mechanism. The procurement strategy which uses the lowest price award mechanism is known as lowest price-based procurement; the strategy that uses the EMAT award mechanism is known as value-based procurement.

The evaluation technique in the lowest price award mechanism simply consists of rejecting bids that do not comply with the Table of Requirements (ToR) and selecting the cheapest bid. In reality, this rarely occurs, because bids almost never comply entirely with the ToR or other contractual provisions – there is usually some qualification.

The EMAT award mechanism takes into account other criteria than price and conformance with requirements. These other criteria are known as award criteria and are used to establish the partial performance of each bid. The evaluation technique combines the performance and price information into a preference ranking. Generally, the evaluation technique uses a mathematical formula.

Legal requirements

Stipulation 46 of the directive (European Parliament, 2004) states:

> Contracts should be awarded on the basis of objective criteria which ensure compliance with the principles of transparency, non-

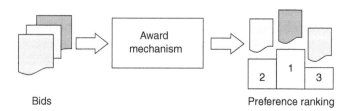

Bids   Preference ranking

Figure 13.1 The award mechanism grades the bids.

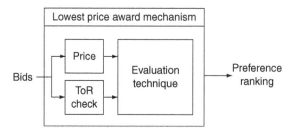

Figure 13.2 The award mechanism in lowest price-based procurement.

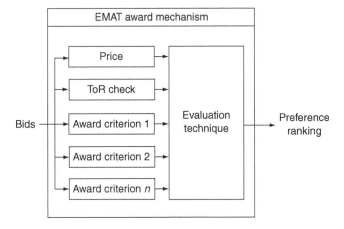

Figure 13.3 The award mechanism in value-based procurement.

discrimination and equal treatment and which guarantee that tenders are assessed in conditions of effective competition. As a result, it is appropriate to allow the application of two award criteria only: 'the lowest price' and 'the most economically advantageous tender'.

Pijnacker Hordijk *et al.* (2004) elaborate these legal requirements: to summarize, the award criteria should be objective, transparent, non-discriminative and should be known in advance by the suppliers.

Types of evaluation technique

Several legally acceptable possibilities for combining financial and non-financial criteria exist (Goovaerts *et al.*, 2004). Up to the present, four types have been identified from the directive and from use in practice.

1   **Price correction.** The performance of a bid on initially non-financial criteria is expressed in monetary terms, which is then combined with the price of the bid. The bid with the lowest price after this correction is awarded the contract.
2   **Point system.** The performances on financial and non-financial criteria of a bid are translated into points. The bid with the most points is awarded the contract.
3   **Ratio system.** The total value of an offer is expressed in a number, which is divided by the price; the offer with the highest ratio wins.
4   **Design contest.** Bids are evaluated on non-financial criteria only, because they all have to comply with the same budget and other contractual demands.

By choosing between these techniques and shaping the chosen option, the public client establishes the influence of financial and non-financial criteria in the assessment.

Research methodology

In order to raise the confidence level of public clients in value-based procurement, knowledge about this approach has to increase. An earlier section explained the most important element of value-based procurement, namely the EMAT award mechanism. The most distinctive elements of this mechanism are the award criteria and the evaluation technique. These elements have to be studied in order to increase knowledge about the EMAT award mechanism. An important methodological consideration is that information is gathered via realized applications, because this information cannot be influenced by the study. Information is gathered, first, by archive research and, second, through interviews. For the collection of information, a template is used which is developed via an iterative procedure. The study is limited to cases from the Dutch construction industry and to award criteria only (no selection criteria). Unsuccessful cases are also included. The administrative procedure surrounding the award mechanism is, however, outside the scope of this study.

Case studies

This research defines a case as a project wherein the EMAT award criterion has been adopted by a public client. In this section, cases are presented individually and in aggregated form.

Case outline descriptions

Case A – design and construction of a secondary school

Budget: about €20M

Type of evaluation: price correction

Evaluation formula: $Price_{Base} + Price_{Maintenance} + Price_{Scenarios} - S_{Energy} - C_{Wishes} - C_{Cooperation} - C_{Value\ creation}$

Award criteria: see Table 13.1

Table 13.1 Award criteria used in case A

Award criteria	Range (€)
Energy performance	[0–1,800,000]
Satisfaction of wishes	[0–2,000,000]
Level of cooperation	[0–500,000]
Value creation	[0–1,000,000]

Case B – design for the renovation of a large office building

Budget:	about €25.5M
Type of evaluation:	point system
Evaluation formula:	$\text{Points}_{\text{Total}} = 0.3 * \text{Points}_{\text{Price}} + 0.7 * \text{Points}_{\text{Award criteria}}$
$\text{Points}_{\text{Price}}$:	$(2.5 * \text{Price}_{\text{Lowest bid}} - \text{Price}_{\text{Bid x}})/(1.5 * \text{Price}_{\text{Lowest bid}}) * 5$ points, with a minimum of 0 points.
$\text{Points}_{\text{Award criteria}}$:	see Table 13.2. Maximum number of points to be earned with these characteristics: 5.

Table 13.2 Award criteria used in case B

Award criteria	Weight (%)	Range (points)
Aesthetics	20	[1–5]
Quality of office concept	10	[1–5]
Preservation of monument	10	[1–5]
Quality of employees	10	[1–5]
Cooperation	15	[1–5]
Product quality	15	[1–5]
Explanation price	20	[1–5]

Case C – design and renovation of a large office building

Budget:	about €190M
Type of evaluation:	point system
Evaluation formula:	$\text{Points}_{\text{Total}} = \text{Points}_{\text{Price}} + \text{Points}_{\text{Award criteria}}$
$\text{Points}_{\text{Price}}$:	the lowest bid attracts 100 points; other bids attract 1 point less per € million difference, with a minimum of 0 points.
$\text{Points}_{\text{Award criteria}}$:	see Table 13.3. Maximum number of points to be earned with these characteristics: 30.

Table 13.3 Award criteria used in case C

Award criteria	Weight (%)	Range
Visual quality	40	Unknown
Functionality	40	Unknown
Flexibility	20	Unknown

Case D – design and reconstruction of a road and several features

Budget:	unknown
Type of evaluation:	point system
Evaluation formula:	$\text{Points}_{\text{Total}} = \text{Points}_{\text{Price}} + \text{Points}_{\text{Award criteria}}$
$\text{Points}_{\text{Price}}$:	the bid with the lowest Net Present Value attracts 90

points; the other bids attract 2 points less per percentage point difference.

Points$_{\text{Award criteria}}$:   see Table 13.4. Maximum number of points to be earned with these characteristics: 7.

Table 13.4 Award criteria used in case D

Award criteria	Range (points)
Nuisance surrounding area	[–1,0,1]
Traffic safety	[0,3]
Aesthetics	[–3,0,3]

Case E – design and construction of a quayside wall

Budget: about €60M

Type of evaluation: point system

Evaluation formula: Points$_{\text{Total}}$ = Points$_{\text{Price}}$ + Points$_{\text{Award criteria}}$

Points$_{\text{Price}}$:   the lowest bid attracts 120 points; other bids attract 1 point less per percentage point difference, with an unknown minimum.

Points$_{\text{Award criteria}}$:   see Table 13.5. Maximum number of points to be earned with these characteristics: 80.

Table 13.5 Award criteria used in case E

Award criteria	Range (points)
Planning	[0–16]
Risk	[0–16]
Maintenance	[0–16]
Innovation	[0–16]
Quality	[0–16]

Preliminary database

Table 13.6 presents the main characteristics of the evaluation techniques. The first column names the cases. The types of evaluation techniques in the second column correspond to two of the four types mentioned earlier. The column 'Share award criteria' presents the possible influences that preferences have on the performance of award criteria. The column 'Performance determination' states how the performance of award criteria was determined. Using the relative method, knowledge of the other bids is needed, because all bids are ranked per award criterion. The absolute method uses an absolute performance standard; so using that method, knowledge of other bids is not required.

Table 13.6 Main characteristics of the evaluation techniques used

Case	Type of evaluation	Share award criteria (%)	Performance determination
A. Secondary school	Price correction	18*	Absolute
B. Office building	Point system	70	Relative
C. Office building	Point system	23	Relative
D. Road and features	Point system	7	Relative
E. Quayside wall	Point system	40	Relative

Note
*Approximation based on cost estimate. This share depends on the price of the offers.

Systematic considerations

The interviewees directly and indirectly confirm the research assumptions in the problem statement. They support the idea of increasing the level of value-based procurement and stimulating supplier creativity. Even in cases where they already have knowledge about setting up value-based procurement, they are still interested in this study for several reasons: they assume it can help them to do it better and quicker next time; it can confirm they have done the right thing; it can help build up confidence with regard to legal aspects; and finally, they feel it can help them explain their decisions to controlling authorities. Moreover, they believe that a thorough, structured overview of the strengths and weaknesses of the possible configuration options in EMAT award mechanisms will be a valuable addition to their toolkit and professional knowledge. They state that a value-based procurement strategy has led to a higher value–price ratio than would have been likely with a traditional procurement strategy.

Point system preferences mapped into the value–price model

The value–price model is a notion of value for money; consumers do not look at the price of a prospect alone, but at what it will deliver. According to this theory, customers will choose the product with the best value–price combination. In the example in Figure 13.4, prospect number 1 is preferred over prospect number 2, even though it is more expensive.

There are two preference systems: one that bases preference on 'value minus price' and one that bases preference on 'value divided by price'. The example in Figure 13.4 is of the latter type, as can be discerned from the diagonal lines of equal preference. A line of equal preference is a collection of points for which an equal preference exists. In a 'value minus price' preference system these lines are parallel. One could say that a 'value minus price' preference system evaluates offers on the procurement profit; whereas, the 'value divided by price' preference system evaluates offers on (societal) efficiency.

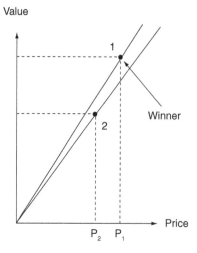

Figure 13.4 The value-price model and two lines of equal preference.

There is a need to visualize point system preferences within the value–price model. As the value–price model has two dimensions, it can provide a more insightful picture of how offers relate to each other than would be possible from a one-dimensional point ranking. Furthermore, the implications of using a point system can be explored. For this purpose, the point system of case E is combined with three fictitious offers (Table 13.7).

This section uses the mechanism of case E as example to illustrate how the results of a point system award can be presented in the value price framework. The bid with the lowest price attracts 120 points; other bids attract 1 point less per percentage price difference. Performance over five criteria can lead to a maximum added value of 80 points. The offer with most points wins. Table 13.7 presents a fictitious bid situation.

Table 13.7 Fictitious bids for the point system of case E

Offer	Price (M€)	ΔPrice (%)	Pts.$_P$	Pts.$_Q$	Pts.$_T$	Rank	Added Value (M€)	Price-Added Value (M€)
A	60	–	120	30	150	3	€0.6M/pt.* 30pts.= 18	42
B	70	16.67	103.33	50	153.33	2	€0.6M/pt.* 50pts.= 30	40
C	80	33.33	86.67	70	156.67	1	€0.6M/pt.* 70pts.= 42	38

Notes
Pts.$_T$ = total points, Pts.$_Q$ = points; for quality; Pts.$_P$ = points for price.

The first step for presenting the results in the value–price model is to acknowledge that the point system corresponds with a 'value minus price' preference system instead of a 'value divided by price' preference system and that the number of points corresponds to the 'equal preference' lines – the higher, the better. Instead of value, the vertical axis corresponds with added value since only the added value is known. The second step is to establish the reference line; in this case the bid with the lowest price attracts 120 points (see the 120 points line in Figure 13.5).

By using the statement that 'the other bids attract 1 point less per percentage price difference' the value per point can be calculated. In this case, 1 point = 1 percent price difference of €0.6M. Based on this information, the other 'equal preference' lines can be drawn. It then becomes clear that the lines through offers B and C are the 103.33 and 86.67 point lines respectively. The added value can be calculated in two ways: graphically via following the point lines, or mathematically by simply calculating the monetary value of the added value points. Once that value is known, the 'corrected price' can be calculated too – see the column 'price minus added value' in Table 13.7. These corrected prices are represented in Figure 13.5 by P'.

Conclusions

The example in the previous section shows that the point system is of the 'value minus price' preference system, instead of the 'value divided by price' system. Trying to map the results of the point system to the 'value divided by price' preference system is likely to yield strange results. Furthermore, it should be clear that point systems operate according to the same principle as the price correction mechanism, but obscure that fact by

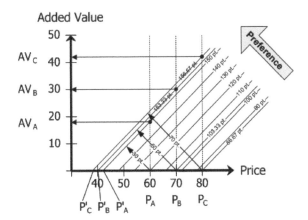

Figure 13.5 Point system presented in the value price framework.

first translating the price into points. In order to obtain an overview of what is really happening, one would have to translate the point score back into monetary terms. It is advisable to avoid this confusing double translation and just use the price correction mechanism. From the previous section it can be concluded that the value–price model is suitable for modeling EMAT award mechanisms and presenting the bids in such a way that their relative performance can be assessed quickly.

The construction industry is often divided into residential, commercial and civil sectors. Until now, no EMAT award mechanisms have been encountered in the residential sector. The number of cases for the commercial and civil sectors is about equal. Looking at the column 'Share award criteria' in Table 13.6, the possibility to influence preference by adding value seems a little higher in the commercial sector than in the civil sector. Based on the five cases, public clients seem to prefer point systems over price correction systems. This conflicts with the findings of the previous section and statements made by some experts that public clients prefer price correction systems, because they are more intuitive, easier to explain to controlling authorities and easier to formulate.

Notes

1 Article 53.1 of *Directive 2004/18/EC* of the European Parliament and of the Council of 31 March 2004.
2 www.ogc.gov.uk/procurement_the_bigger_picture.asp.

14 Organizational culture in the merger of construction companies

Karin Johansson

Introduction

Mergers and acquisitions in the construction industry are fairly common, particularly as economic constraints make it harder for mid-size construction companies to grow organically into larger companies. In addition, mergers can be seen as part of a strategy designed to achieve growth. Initially, when merging two companies, employees within both organizations might experience confusion, stress or even fear due to the uncertainty they experience in the new situation. On the organizational level, these experiences can be shown in lowered commitment and productivity, as well as in increased dissatisfaction and disloyalty (Buono and Bowditch, 1989). Indeed, many companies seem to overestimate the potential benefits from merging two companies (e.g. an increase in market share) and the ease with which the merger can be accomplished. Studies show that most failures originate from lack of communication or differences in management styles and culture. In fact, it has been confirmed that as much as 80 percent of the risks associated with mergers and acquisitions originate from poorly managed cultural integration (Brahy, 2006).

This chapter reports on a study of the merger of two mid-sized construction companies in the Gothenburg region of Sweden. The companies were considered to be a good match by both parties since they operate within different, but complementary, sectors within the construction market. When merging, however, other factors than optimized joint activity must be considered. These factors, such as differences in culture and styles of management, have an impact on whether a merger will be successful or not. While studying the merger between the two construction companies (A and B), questions arose on how the merger could become as efficient as possible. The most distinctive difference between the two companies is that company A focuses mainly on its projects and financial position in order to achieve profitability, while company B places a higher value on its organization and employees' personal development.

Since cultural differences tend to be overlooked during mergers, it is one of the drivers for this study. In this connection, communication can be regarded as an essential tool (Doherty, 1988). The aim of this study is to outline how these two construction companies can manage cultural differences using different means of communication.

Background

Ten years ago, a large construction group started to develop its interests in the western region of Sweden by founding company A, together with employees attracted from an existing company in the region. Over the years, company A has been successful in experiencing continuous market growth. Company B was founded in 1981 and has experienced a similar development path to company A. Both had looked at different ways to expand without raising their overhead costs and saw merging with another company as the best way forward.

Company A was owned and managed by a majority of its employees and parent company – the large construction group. In May 2006, the construction group acquired company B with the intention of merging it into its operations in March 2007. Since the construction group was a joint-stock business, employees within the companies could not be informed about the acquisition or merger before the official announcement.

In order to find ways to enlarge the company, company B undertook an internal investigation. As a result of the investigation thoughts of an acquisition arose. The investigation showed that there were, however, many barriers to overcome before such a change could be made and that it could lead to unforeseen problems.

The acquisition of company B was a strategic move by the large construction group, since the action would gain the group market share as well as increase its presence in the region. The reason for acquiring company B was that the combined organization, which a merger of company A and company B would create, could enable the group to offer a greater range of competences to the market. An important issue for company B during the bid and negotiation process was that its organizational values should be shared by the acquiring company. Only top managers within the two companies knew about the ongoing process and negotiated the conditions as they saw fit. The acquisition and merger needed to be kept confidential as the large construction group is listed on the stock market. The news was broken to employees at meetings arranged for the same day as the acquisition became public.

The study, which this chapter discusses, was based on literature covering cooperative culture, mergers, acquisitions and communications. Additionally, seven semi-structured and individual interviews with employees of the two companies were undertaken three months prior to the actual merger. The individual interviews were guided by open-ended questions in order to encourage interviewees to speak more freely about their thoughts and feelings regarding the merger. Questions were formulated to outline how and to what extent communication was used to circulate information throughout the organization and to reveal any informal communication patterns. The interviewees were selected in order to gain different insights into how communication was carried out within the two companies and were drawn from various positions within them: two CEOs, a CIO, cost estimator, quality and environmental manager and the head of civil engineering. Thus, managers in charge of strategic information, as well as employees receiving this kind of information, were interviewed.

Literature review

There are several definitions of organizational culture of which Schein's (1985) was used for the purpose of this work.

> The pattern of basic assumptions that a given group has invented, discovered, or developed in learning to cope with its problems of external integration, and that have worked well enough to be considered valid and, therefore, to be taught to new members as the correct way to perceive, think, and feel in relation to these problems.
>
> (Schein, 1985: 6)

Doherty (1988) offers ways of acting to resolve cultural conflicts based on experience from eight mergers. Moreover, he declares that it is essential to establish quickly who is in charge; subsequently, it will be easier to resolve practical issues. Resolving issues at an early stage serves to avoid anxiety and rumor. Doherty emphasizes the importance of choosing a leader who believes in the merger and can convince employees as well as clients. In addition, the companies must take into consideration the individual employee and, if possible, arrange one-to-one meetings between managers and employees. The gain from direct communication is that the anxiety created by uncertainty can be reduced. Indeed, reducing anxiety is the aim since any concern shines through when the employee is in contact with clients.

Affect of organizational structure on communication

Well-performed communication leads to fewer problems within an organization (Jacobsen and Thorsvik, 2006). Information tends to be channeled to top managers, who disseminate what they find to be relevant for their employees; thus, information flows in one direction. Jacobsen and Thorsvik argue that alternatives for employees to inform their managers have to be established too. According to Hall (1972), the more levels there are within an organization, the more information tends to be left out on the way. Mintzberg (1979) contends, however, that such a situation does not necessarily have to be negative. Information that is less important or poorly augmented does not reach top managers and so they can focus on what is most relevant.

Jacobsen and Thorsvik (2006) found that information intended for horizontal circulation in an organization is not guaranteed to reach all concerned. In fact, information decreases considerably when crossing an organizational boundary (Egeberg, 1984). Hence, when an organization stays small, informal communication may be enough and no structural communication strategy or tools are needed. In situations where organizations are expanding or merging, informal ways of communicating will not satisfy the need for exchanging information. The demand for structured communication grows as the organization grows (Egeberg, 1984).

Organizational structure has three general effects on behavior: stability, restriction and coordination (Jacobsen and Thorsvik, 2006). A position within the organization comes with certain assignments and routines to solve them, which creates stability and places restrictions on what can be worked on. Moreover, the coordination of different assignments may lead to better performance of the organization overall than is the case where each individual is working on his/her own (Jacobsen and Thorsvik, 2006). This view is in line with previous studies on team effectiveness and, as pointed out by Johnson and Johnson (2006), working in teams results in higher individual productivity than working competitively or individualistically. A functionally based organizational structure means that similar working tasks or assignments are brought together in the same organizational unit. According to Jacobsen and Thorsvik (2006), this way of organizing an organization has both advantages and disadvantages. The most important advantages are maximum specialization based on similar assignments and the avoidance of duplicated work. As the main disadvantages, Jacobsen and Thorsvik (2006) mention lack of interest in, and understanding of, others' work as well as coordination problems between departments and functional units.

Importance of organizational culture and communication

Factors contributing to a common organizational cultural spirit are artifacts, standards, values, views and assumptions (Schein, 1985; Hatch, 2001). Artifacts are concrete physical objects and ways of acting that express underlying norms, values and assumptions. Jacobsen and Thorsvik (2006) regard artifacts as symbols that have to be interpreted and as intermediaries for information and organizational culture. Yet, artifacts have more than symbolic value; they can be used to solve problems.

Collins and Porras (1994) reveal a strong connection between a solid organizational culture and an organization's success. Companies with a united organizational culture have concrete methods for socialization, leading to employees sharing the company's core values. Kotter and

Heskett (1992) likewise acknowledge that a strong organizational culture is connected with the performance of an organization. Mayo (1945) studied subcultures and declared that it was difficult to rule a group with formal means when the group had strong values of its own. As later confirmed by Kotter and Heskett (1992), the achievement of the group is connected with the organization's culture. Hence, organizational culture affects the way that employees communicate with each other. In organizations where cultural difference is a major feature, communication has to be forced in order to grow into a natural part of the organization. Communication may be used to create a social situation; this resulting social bond will then serve as a basis for the trust that is needed for communication to work properly (Jacobsen and Thorsvik, 2006).

Research results

Overall position

The general opinion among top managers within both companies was that the merger would prove advantageous for their clients. Even so, some employees had a different opinion on the merger and found it hard to articulate the advantages with clients, although the prevailing view was that the new organization would have a wider and better range of experiences and competences. All interviewees believed the merger would lead to improved competitiveness in the market. For the merger to be truly successful, the interviewees felt it was necessary for top managers to be clear about the goal and mission of the new company.

Company A

Interviewees felt that the company had reached its organizational limit and the preferred way to achieve additional growth would be by merging with a construction company of a similar size. For this reason, the forthcoming merger with company B had been met with optimism by the majority of employees. Nonetheless, one interviewee maintained that the two construction companies had different concerns over the merger and this was borne out from discussions in other interviews. Employees from company A considered that company B's employees were more concerned about the merger than themselves. As noted earlier, company A had been owned and managed by a majority of the employees since it was founded. Communication within the company had mostly been conducted through meetings with top managers and leaders at the project level. The outcomes from these meetings were, according to the interviewees, distributed throughout the company by email. Work conducted within the company was reviewed in the internal newsletter which was distributed to all employees.

The merger process, where the construction group would be taking full ownership of company A might, according to the interviews, create an unusual situation for employees. Interviewees felt that the merger affected the ability to exert influence over the organization's progress, in terms of strategic matters and management control. During the interviews carried out at company A, ability to influence daily work was regarded as of great importance by employees. Moreover, this had created a united spirit among employees as decisions were supported across the organization. From the interviewees' perspective, it was important to continue with meetings to retain the ability to influence the new organization.

Further concerns were expressed: employees of company A sensed a significant difference in the way the two companies carried out their work. 'We have a large pool of staff and workers for all our three working groups, whereas company B is working more closely together in small teams which function almost like small companies' (Employee, company A). Interviewees believe the new organization has to function in a similar way to be able to continue to secure contracts from a wide range of clients. They also think that a proportion of employees (staff and workers) would have to change position within the new organization in order to create a common organizational culture.

Company B

Thoughts about an acquisition first arose when company B undertook an internal investigation into how to develop and expand its business. Having abandoned all other ideas in favor of primarily seeking a merger, the company's top managers entered into negotiations with company A. Before the acquisition and merger became public, company B experienced a lot of rumor-mongering in its corridors. As one interviewee put it 'a lot of things were put on ice without any explanation. Today, information is given more informally, for example at the coffee table or in the corridor' (Employee, company B). Aside from rumor-mongering, interviewees spoke first about the surprise they experienced, and later the feeling of anxiety, as questions about their continued function within the new construction company were raised. The information given at the first meeting consisted mainly of, what looked like, a perfect match between two mid-sized construction companies. At this first information meeting, interviewees felt that little attention was given to the human side of the merger, such as employees' feelings and questions about the future. The information provided made clear that everyone within the two companies would have a place in the new organization, but not their position.

Most interviewees stated that the pipelines used after the announcement to communicate within the organization were no longer sufficient to ease employee concerns over the merger. During the interviews with employees at company B, it became clear that they wanted more information on an individual basis regarding their function within the new organization. In addition, ideas on how to circulate information throughout the new organization were offered, e.g. proper use of the intranet and emails with information to all employees. They added that with communication it would be possible to complement each other and share experiences across the former companies. According to the interviewees, they wanted clear and early formal information to avoid rumor-mongering, which they said could create a bad atmosphere. Furthermore, different information needs are crucial for a successful merger and to avoid a 'them and us' mentality taking hold. 'It is of enormous importance to communicate to make the different cultures fit together' (Employee, company B). Following the public announcement, employees in company B expressed their anxieties about the differences in management culture between the two companies. Many employees in company B were content with the familiar atmosphere and were keen to keep this in the new organization. Another view was that they wanted to retain a mid-size company spirit, symbolized by standing up for the individual. Top managers must keep the respect and maintain contact with individuals so they feel a part of the future.

The new organization was considered to be in need of a common way of communicating with clients and employees in order to make them comfortable and believe in the new organization; for example: 'We need a new common logotype, not three as we're going to use. Neither client nor employees will experience continuity if they find three different signs everywhere' (Employee, company B).

Discussion

The literature makes clear that communication can be used as a tool in the merger process between companies having different organizational cultures. When forming a new organization, Jacobsen and Thorsvik (2006) suggest several conditions or actions which can increase the possibility of creating a common culture. Of these, the need to make a concerted effort to circulate information throughout the organization is paramount. As the organization grows, the more levels it has and the more important it is to use clear pipelines (Hall, 1972). When two companies merge top managers must focus unambiguously on employees. By circulating information early and in a formal and correct way, employees may well experience a more harmonious merger. This is clearly of major importance in achieving an effective merger.

The need for information and communication has been confirmed as an important factor in bringing about a successful merger. However, the interviews reveal that top managers had not necessarily reflected on this matter. The interviews show that, at an early stage in the merger, lack of information made employees insecure. Anxious employees will become less effective which, in the long run, can affect day-to-day business. By communicating to a greater extent, relevant information will reach employees and in doing so will enable them to focus on their work.

Creating a common organizational culture within the new organization is of great importance as confirmed by Collins and Porras (1994). In their study, there is a direct connection between organizational culture and success. A successful organization has concrete methods for helping employees share the company's organizational culture. Although our interviews show that no clear strategy has been devised on how to achieve a common culture, interviewees with high positions within both companies state that it is important to avoid a 'them and us' mentality. From the study, it seems that most work prior to the merger focused on functional and practical issues. Pre-existing groups of employees in both companies were retained as far as possible in the new organization. In this case, and in accordance with Collins and Porras (1994), it would have been advisable to consider a greater mix of employees from the two companies in project organizations rather than keeping existing organizational structures.

Formal communication between managers and employees can avoid uncertainty and rumors. It is necessary for employees to have the chance to create a positive opinion of the other company. There are several indications in both companies that their own culture is unlike the other. Where this is the case, formal communication must be used to handle the differences at an early stage.

Many aspects must be taken into consideration to create a common culture of cooperation. The general opinion of top managers is that a physical object such as a logotype is not essential to engendering a common organizational culture and spirit. After the merger the organization will use a combined logotype, consisting of the two originals. Even so, a clear and distinct message to the market may be a good start for the new organization. In line with the views of Schein (1985) and Jacobsen and Thorsvik (2006), a new corporate logotype would be the first step toward a cooperative culture. It is important to acquaint clients and the market with the new company culture. In this case, both interviewees and previous research (Schein, 1985; Hatch, 2001; Jacobsen and Thorsvik, 2006) emphasized the need for a common artifact, such as developing a new logotype. Keeping a logotype of combined original appearance may initially make clients secure; but in the long run, a newly developed logotype will indicate a stronger, merged organization, which is important for clients as well as for employees.

Ways for employees to inform their managers have to be established as well as managers informing their employees. Company A's way of disseminating information by regular meetings would help to implement the common culture, something that company A wants to develop. Furthermore, both companies use an internal newsletter which has to be developed into a common pipeline. By developing this way of informing employees within the new organization, an opportunity is created to convey the organization's core values.

Conclusions

The results of the study were intended to assist the two merging construction companies in handling communication in a way that would benefit their emerging common organizational culture. As the merger of these two companies was considered to be a good match, the owners and top managers tended to focus more on the function and practical issues of the merger than on employees' opinions. As the results show, employees experienced uncertainty for the future, partly due to lack of information. It is our recommendation, therefore, that top managers carry out a structured information process with all employees in order to bring forward any misunderstandings regarding their future employment and to provide information concerning their new role in the new organization.

It is suggested that the new organization introduces a common logotype to reflect and contribute to the new organizational cultural spirit. In addition, the merged company needs a new, organized system for meetings to ensure dissemination of information. Moreover, it is suggested that a joint intranet be developed which employees can use to access necessary information. In order to circulate information to the organization as a whole, and to create a common culture for all employees including workers in the field, we recommend a regular, written report from each site and department. All contributions can then be summarized and published in the new common company newsletter.

Acknowledgment

The author would like to thank Daniella Balogh, Mikael Hultqvist, Jon Jansson and Tomas Sandahl for contributing to the empirical research used in this chapter.

15 Environmental attitudes, management and performance

*Pernilla Gluch, Birgit Brunklaus,
Karin Johansson, Örjan Lundberg,
Ann-Charlotte Stenberg and
Liane Thuvander*

Introduction

Over the past two decades the Swedish construction industry has made a lot of effort to develop green building practices. Researchers within the field have provided a theoretical understanding of how to design green buildings and analytical environmental management tools have been developed to guide practitioners. Furthermore, information campaigns have raised general environmental awareness amongst practitioners. In spite of these efforts, mainstream building practices do not seem to have undergone any marked changes (Gluch, 2005; Femenías, 2004). Progress toward a viable and sustainable construction industry relies on its ability to foster and transfer innovative products, services and practices (Keast and Hampson, 2007). However, the absence of the large-scale innovation necessary to drive this development forward is evidence of an imperfect process. This raises a number of questions: why is it so difficult to incorporate environmental issues into mainstream business? How are environmental issues actually dealt with in the construction industry? Has development stagnated? What is causing green innovation inertia in the industry? Fundamentally, *what makes it slow?*

This chapter aims to provide some answers to these questions by empirically examining environmental attitudes, management and performance in the Swedish construction industry. The chapter is based on a questionnaire survey carried out in the autumn of 2006 which is almost identical to one carried out in 2002 (Baumann *et al.*, 2003). The questionnaires were sent to environmental managers or their equivalent in firms having at least 50 employees in real estate, engineering and construction, and architectural firms with at least 20 employees. This covered 542 firms and resulted in a response rate of 45.4 percent. The structure of the survey included the industry's definition of environmental challenge, attitudes toward the challenge and the performance of, and response to, environmental measures taken by the firms.

Results from the 2002 study showed that many firms at the time were working with environmental issues. However, the study showed that their work focused mainly on a few targeted areas, e.g. toxic substances and waste management, which departed from what they perceived as the industry's main challenge – energy savings. Firms placed much emphasis on high-level environmental management activities, e.g. environmental management systems (EMS), while the implementation of technical environmental measures met with considerable resistance. Of particular note was that significant focus was placed on pre-planning activities while feedback and self-assessment were neglected. This resulted in asymmetric communication within the firm, with the consequence that many environmental managers lacked information about their firms' environmental performance. By repeating the survey, it has been possible to identify trends and institutionalizing processes that contribute to, as well as hinder, sustainable development and green innovation within the construction industry. This chapter points toward some possible explanations as to why the development of environmental measures sometimes does not go in the direction intended by senior managers despite receiving attention and effort.

Research method

The 'Environmental Barometer for the Construction Sector' is a questionnaire-based study with the objective of surveying environmental attitudes, management and performance within the Swedish real estate and construction industry. The structure, as schematically illustrated in Figure 15.1, has been developed from the questionnaire used by the 'International Business Environmental Barometer' (IBEB), which has measured the state of environmental management in the industry since 1993. The terminology and wording in IBEB's standardized questionnaire has been adjusted to take account of terms and words better suited to construction. The structure of the survey covers the industry's definition of its environmental challenge, attitudes toward this challenge and the performance of, and response to, environmental measures.

Figure 15.1 General structure of the survey.

Preparation of the questionnaire

The questionnaire used in the 2006 study has changed just slightly from that used in the 2002 study. A deliberate intention was to keep the questionnaire as intact as possible in order to be able to make comparisons over time. The 14-page questionnaire contains five main sections: business characteristics, environmental management, environmental impact, environmental measures and reflections on the effects of measures taken. A section directed solely at real estate firms concerning energy declarations was also added, but is not presented in this chapter. The questionnaire contains a total of 39 main questions, most of which have alternative sub-questions.

Statistical population

The survey covers all companies in Sweden with at least 50 employees within construction (NACE group code 45, executing construction companies), development (NACE group code 70, property owners and managers), consulting engineering (NACE group code 74202) and companies with at least 20 employees within architecture (NACE group code 74201). At the time of the survey, 620 companies had a core business that fell into one of these categories. Several of the companies, especially consulting engineers, do not, however, belong to the construction industry, for example ICT and energy consultants. After correction, the final population to which the questionnaire was sent consisted of 542 companies and/or other organizations. The questionnaires were directed at environmental managers or their equivalent.

Organization of survey

The questionnaire, under cover of an introductory letter, was sent to each company: their addresses were obtained from the companies' register at Statistics Sweden. Three reminders were sent out, the last of which contained a copy of the questionnaire. In addition, and for the purpose of investigating reasons for non-response, an email was sent to companies that had not returned their completed questionnaire after the second reminder. Data were abstracted from the questionnaires and analyzed using the statistical program, SPSS. In order to secure reliability and validity of the survey, a statistician was consulted before data collection and after the analysis had been performed.

Validity and reliability of the study

There is always a risk in surveys intended to measure peoples' attitudes and values that respondents will answer as they believe they should answer and/or attempt to place themselves and their companies in a (more) favorable light. It is, therefore, important to acknowledge that the survey does not present an objective truth about the companies' environmental work, but rather it measures what respondents perceive as their environmental challenge, problems and so forth. There is also a further risk, since the survey is directed to environmental managers; they generally have a larger interest in environmental issues and, therefore, may not be representative of the overall values prevailing within the companies.

Moreover, it may be the case that companies paying more attention to environmental management will be more benign when answering, which might lead to results that are unrepresentative of the construction industry as a whole. However, pre-testing the questionnaire on practitioners, having an informative covering letter with detailed contact information in case of queries, sending multiple reminders and investigating the reasons why some respondents failed to respond helped to reduce bias in the result caused by problems of interpretation and non-response and is in line with recommended research practice (Bryman, 2008).

The significance of this discussion is that the reader can recognize the potential for bias in the interpretation of results. This chapter presents basic frequency analysis only, whereas the database permits more advanced and detailed analyses which would strengthen the study's validity. A more detailed description of the methodological approach can be found in Gluch *et al.* (2006).

Results of the survey

246 environmental managers out of a possible total of 542 completed the questionnaire, corresponding to a response rate of 45.4 percent. The distribution of the four groups is presented in Table 15.1.

Environmental challenge as perceived by the companies

The environmental challenge is defined by how the companies see themselves contributing to environmental problems and how they experience environmental pressure from stakeholders. The following sections discuss these findings.

Table 15.1 Total number of companies, response frequencies and rates

	Sample size	*Rate(%)*	*Responses*	*% Rate*	*Answers (%)*
Construction companies	300	55.4	123	50.0	41.0
Real estate firms	151	27.8	78	31.7	51.7
Architectural firms	36	6.6	20	8.1	55.6
Consulting engineers	55	10.2	25	10.2	45.5
Totals	542	100	246	100	45.4

Environmental problems

Most companies see the use of non-renewable resources, energy and water as their most serious environmental problems. These three areas were also where a majority of the respondents perceived they had lowered their impact. They see their least serious problems in the areas of contaminated soil, risk of environmental accidents, waste management and use of toxic substances (see Figure 15.2). Energy aspects, global climate change and waste were put forward as the construction industry's major challenges now and into the future.

Stakeholder pressure

Customers/clients and managers are regarded as the most environmentally influential stakeholders by most companies (see Figure 15.3). The final customer is also considered an important stakeholder along with employees and owners/shareholders of the company. When seen from an environmental research perspective, as well as from one of environmental information, it is noticeable how little influence researchers, environmental organizations, mass media and politicians are assumed to have on the companies' environmental activities. Neither financial interests, such as banks, insurance companies and financial analysts, nor controlling interests such as those of management accountants are perceived as influential on the companies' environmental activities. There are some differences between the groups within the industry, although the client is placed as primary stakeholder by all.

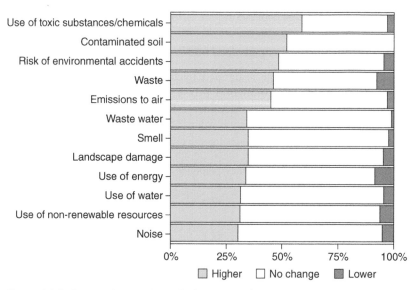

Figure 15.2 Companies' rating of their contribution to various environmental problems relative to the industry average.

Contractors and consulting engineers rank clients higher than developers and architects; developers regard managers and environmental authorities as having a high level of influence.

Companies' responses to the environmental challenge

Companies' responses to environmental challenges can take different forms; for instance, employing specialist personnel and creating environmental working groups, cooperation with other stakeholders, technical measures and managerial measures.

Staffing and environmental personnel

A majority of the companies have personnel for handling environmental issues within the company (81 percent), although the proportion that does not (19 percent) is, compared to manufacturing industry, relatively high (10 percent in 2001 according to Nilsson and Hellström (2001)). Many of the

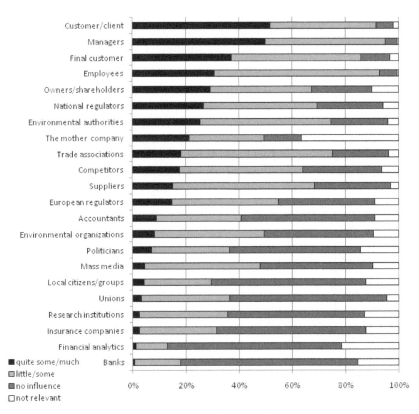

Figure 15.3 Companies' rating of stakeholders' influence on environmental activities in the company.

personnel working on environmental issues do it on a part-time basis, i.e. the person has other tasks besides those of an environmental nature. Most respondents indicated that the number of environmental personnel has remained the same over the past four years (see Figure 15.4). In 2002 (Baumann *et al.*, 2003), the number of environmental personnel was increasing moderately-to-much in the companies, indicating that the number of personnel has stabilized at the level of one person on average per company.

How influential environmental work is in the company is due partly to the official position that the environmental manager occupies in the company. The study shows that a majority of environmental managers (66 percent) are not members of the company board, representing a decrease from the 2002 study when 56 percent were members. There is a difference between the groups such that it is more likely the environmental manager is a member of the board of construction companies (44 percent) than of real estate companies (21 percent).

A majority of the respondents think they have, at least partly, enough knowledge in order to influence practice (85 percent), as well as strategic decisions (85 percent). On the other hand, a relatively significant proportion of the respondents (approximately 25 percent) are not in a position of authority to stop environmentally damaging processes and/or to influence strategic decisions. This reveals a certain discrepancy between ability to influence and authority to do so.

Managerial measures

The environmental activities of many of the companies are undertaken in accordance with an EMS (73 percent). This is a substantial increase since 2002 when 46 percent had an EMS. When combined with companies that are in an implementation phase or are considering implementation of an EMS, the total is 90 percent, thus mirroring the pervasive force EMS has

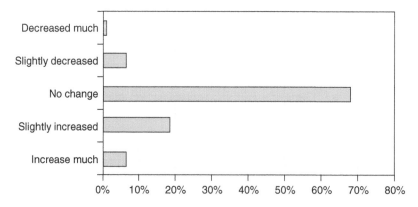

Figure 15.4 Changes in number of environmental personnel during the last four year period.

within the industry. Figure 15.5 shows that managerial activities carried out in the companies are largely related to the EMS; for example, 93 percent of the companies have a written environmental policy, implemented routines to secure the observance of environmental laws (82 percent), established an order of accountability (83 percent) and formulated environmental goals as part of continuous improvement (80 percent) as well as measurable goals (76 percent).

Considering that an overwhelming majority of companies say they set measurable environmental goals, relatively few perform activities that in turn measure environmental performance (see Figure 15.5). Besides activities related to the EMS, the companies foremost carry out activities aimed at transferring environmental information and demands between various actors in the supply chain (see Figure 15.6). Another communicative move is to develop checklists and guidelines. Considering that customers/clients have been put forward as the primary stakeholder, it is surprising that measures such as green marketing and eco-labeling are somewhat rare activities within the companies. In a 'relay' team where many actors are dependent on each other throughout the construction process, from planning to operations, it is surprising that so few are involved in cooperative activities and even more surprising that one-fifth consider it as not relevant.

Cooperative measures

Companies' environmental activities are, just as in 2002, not integrated within the company. Figure 15.7 shows that several areas, such as R&D, accounting, marketing and staff policy have no relation to environmental activities undertaken within the company. Environmental activities have

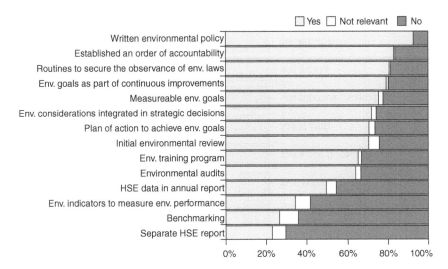

Figure 15.5 Environmental activities related to EMS.

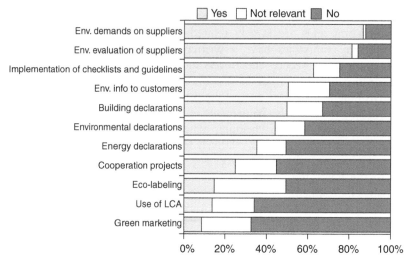

Figure 15.6 Environmental management activities related to purchasing and market.

mostly been integrated with quality, health and safety, which are probably a consequence of companies having organized themselves in this way, with personnel assigned to these multiple tasks.

Most inter-organizational cooperation is carried out with members of the 'classic' relay team, i.e. clients, suppliers and customers (Figure 15.8). The parties with which the companies cooperate also agree about those who they perceive as main stakeholders in their environmental activities

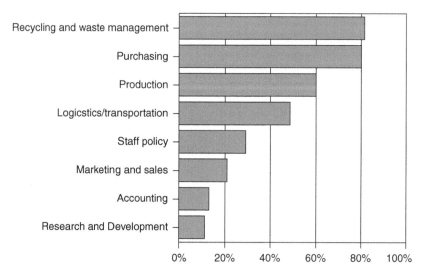

Figure 15.7 Intra-organizational cooperation – business areas where environmental measures occur.

(compare with Figure 15.3). The companies have limited cooperation with R&D units, environmental organizations and departments, accounting or marketing departments (Figures 15.7 and 15.8).

Technical measures

Waste separation is by far the most common measure for reducing environmental impact (see Figure 15.9). Other waste management activities and sub-

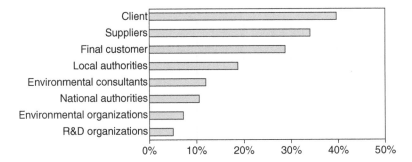

Figure 15.8 Inter-organizational cooperation – stakeholders with whom cooperative environmental measures occur.

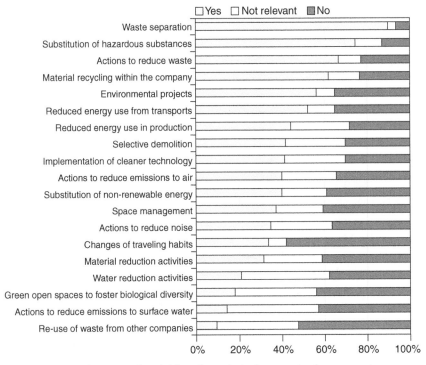

Figure 15.9 Environmental activities of a technical nature in the companies.

stitution of hazardous substances/chemicals are common measures within the industry. Although much effort has been made to reduce waste, several of the respondents regard it as one of the major environmental problems facing the industry. Figure 15.9 also shows that many companies are more devoted to the handling of waste than applying waste minimizing measures. Despite many respondents emphasizing energy as a major problem for the industry, just 39 percent have acted to substitute non-renewable energy sources over the last four years. This is surprising given the importance of energy issues.

Results from the companies' environmental activities

An indication of the success of environmental activities by the companies is found by looking at the impact of those activities on environmental performance and business.

Environmental improvements

Environmental activities have had most impact on waste and use of hazardous substances, non-renewable materials and energy (see Figure 15.10). Apart from energy use, the results are in line with Figure 15.9 which shows that waste management and substitution of hazardous substances are common activities in the industry. The companies point out that in some problem areas there has been no effect or that they have no information about it.

Business effects

In line with the results of the 2002 study (Baumann *et al.*, 2003) and other industry sectors (Nilsson and Hellström, 2001), companies in construction

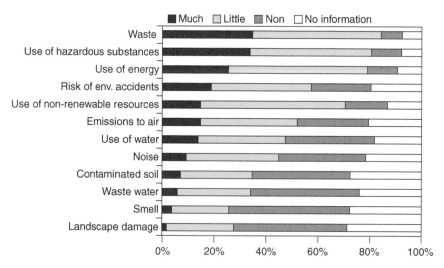

Figure 15.10 Effect of environmental activities on environmental problems.

consider that environmental activities mostly bring long-term benefits to business or benefits for the principal stakeholders, such as staff, management and owners/shareholders. Figure 15.11 shows that a majority of the companies indicated that environmental activities have had a positive impact, especially on company image, whereas they have had a negative impact on profits, cost savings and productivity.

Figure 15.11 also shows that environmental measures taken by most of the companies have had no effect in several business areas. The lack of impact on market factors, such as the creation of new markets and increasing market share, is especially noticeable. This situation can explain the low interest in R&D and innovation, for example clean technology.

Obstacles to effective environmental activities

Obstacles to carrying out effective environmental activities can be divided into internal (Figure 15.12) and external (Figure 15.13), where the latter are out of the company's immediate control and the former are easier for the company to influence. An internal obstacle which many companies emphasize is that of environmental activities proving too costly; they also cite lack of educated personnel. The foremost external obstacle is that of a lack of market incentives. This perception has risen since 2002 and may be the result of respondents experiencing problems entering the green products and services market.

On an overall level, the experience of companies is that obstacles, aside

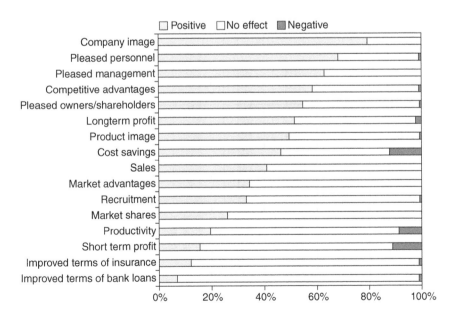

Figure 15.11 Effect of environmental activities on business.

from regulation, are more pronounced now (5–10 percent more) than four years ago. In comparison with other industrial sectors in Sweden (Nilsson and Hellström, 2001), construction regards regulation as a potential solution to its environmental problems.

Discussion and conclusions

From this study, it can be easily concluded that there is an environmental inertia within the construction industry. Although companies within the industry are today active in environmental matters, e.g. having specialist personnel and advanced EMS, the industry overall is struggling. The study shows that the companies' environmental activities still focus on a few targeted measures, the companies continue to have a preference for waste management and environmental activities of a managerial kind and they, as in they did in 2002 (Baumann *et al.*, 2003), consider themselves to have accomplished most results in the areas of toxic substances and waste separation.

We started this chapter by asking, *what makes it slow?* The study reveals five possible reasons for this innovation inertia. First, there is the notion that the market for green products and services is dysfunctional and, therefore, does not stimulate innovation and novel approaches. The lack of market pull for green innovation within the construction industry has also been identified in other countries, for example in the Netherlands (Bossink, 2004). The perception is of an imperfect market and one where environmental work is too costly, making green innovation too risky financially. This belief is also

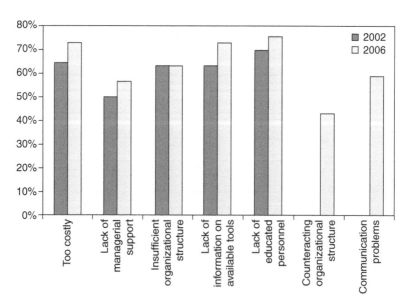

Figure 15.12 Extent to which internal obstacles have influenced environmental activities in the companies.

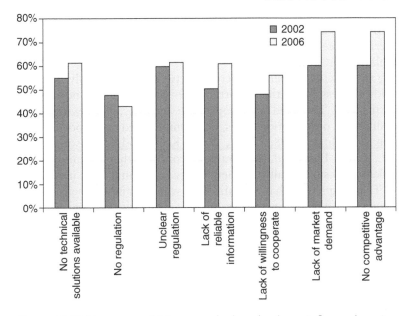

Figure 15.13 Extent to which external obstacles have influenced environmental activities in the companies.

accentuated by the perception that banks and other financial institutions have little or no effect on environmental activities, thus preventing environmental issues being included on the business agenda.

Second, one way of stimulating the creation of a market is through government initiatives in the form of regulations. The present study shows that many environmental managers consider legislation as the most likely solution to the industry's environmental problems. Why legislation – an approach that is usually met with resistance by the industry – is emphasized cannot be discerned from the survey. It might be symptomatic of environmental managers' frustration with getting across their message and so legislation would indirectly strengthen their current weak standing in their companies. It might also be a way for companies to minimize risks by forcing them to be spread over the whole industry. Nevertheless, legislation implacably nurtures bureaucratization and standardization, which is known to restrain the companies' incentive to approach the challenge from a different (and innovating) perspective. Previous research has, for example, shown that regulation may hamper innovation, especially if the regulatory process is too complex and too prescriptive (Gann *et al.*, 1998).

Third, for innovation adoption, it is essential that all actors have enough motivation to support innovative green solutions (Dulaimi *et al.*, 2002) and that they perceive a relative advantage from the new idea (Ling *et al.*, 2007). In order for goals and goal setting to have a motivating effect, it is thus important to provide information on whether one has achieved the goals or not (Locke and Latham, 1984). While many companies say they have set environmental goals, the lack of follow-up activities and environmental performance measurements mean that the motivating effect does not take place.

Fourth, the lack of cooperative actions between actors involved in the construction process limits the opportunity to view products and services in a holistic way. The need for cooperative activities, both within and between different organizations, has been emphasized as important for innovation (Slaughter, 1998; Harty, 2005; Keast and Hampson, 2007; Ling *et al.*, 2007). Ling *et al.*, for example, conclude that for innovations to be implemented successfully a variety of organizational units need to be involved. They emphasize that organizations which maintain their competence through different cooperative means, including internal groups, R&D projects and long-termed relationships with stakeholders, achieve larger innovation capabilities than others.

Last, effective implementation of innovation strategies requires that continuous R&D effort is integrated within the firm's activities (Nam and Tatum, 1992). The present study has shown that companies have limited or even a complete lack of cooperation with R&D departments or institutes, as well as with other environmental knowledge-intense organizations. It can be concluded, therefore, that the foundation and stimuli for the development and creation of pioneering green ideas, innovative green techniques and new green business opportunities are poor within the industry.

We conclude with a successful example of green innovation. In a case study of ten construction projects having a primary project goal to innovate in the field of sustainability (Bossink, 2004), it was found that the increased focus on green innovation not only raised the quality of the projects, but also sustained and enforced the companies' position in the market as well as improved and strengthened cooperative ties and procedures between those involved. So, in order to recommend green business to the industry the answer is not, therefore, to wait for legislation, but rather to be proactive and shoulder the environmental challenge by motivating employees, cooperating more widely and taking appropriate financial risks.

Acknowledgment

The authors gratefully acknowledge the financial support of Centre for the Management of the Built Environment (CMB) at Chalmers University of Technology.

16 Stakeholder management through relationship management

Steve Rowlinson, Martin Tuuli and Tas Koh Yong

Introduction

The conditions under which construction projects are undertaken are conducive to disputes and hostilities from stakeholders. The challenge for the project team becomes one of implementing project strategies such that positive stakeholder's influence is maximized and negative influence is minimized (Walker *et al.*, 2008). Nowhere perhaps is this phenomenon more obvious than in Hong Kong where the populace have, over the past decade, found their voices following the return of the colony to China in 1997 after 150 years of British rule and growing agitation for a more democratic society. The historical context is therefore important in understanding the current situation regarding stakeholder management and relationship management in Hong Kong.

During the colonial years, a British approach to construction was followed, focusing strongly on the traditional approach which was regulated and administered by a strong civil service. This led to a construction industry which relied heavily on hierarchy, tradition and procedures in order to function effectively, but the industry was also heavily influenced by the Chinese culture in which it was situated. Hence, values such as face, harmony and conflict avoidance were also embedded in the industry culture. In such a situation, the issue of stakeholders and their management was paid scant regard; the government was used to making decisions on development rather than consulting widely and the other major players – the oligarchy of large real estate developers – adopted a simple, economic approach to their business plans. Only over the past few years have issues such as corporate social responsibility reached boardrooms. Matters are, however, changing and Hong Kong people have become much more challenging of their government and institutions and have demanded that they be consulted and involved in all developments (e.g. the West Kowloon Cultural Hub, the Tamar Site redevelopment and the demolition of the Star Ferry and Queen's Piers).

In response to this wave of change, major client and construction organizations are embracing corporate social responsibility as a business strategy which in many ways is seen as a driver of stakeholder engagement and management. 'Respect for people' is becoming a core theme in construction organizations. Against this background the issue of relationship management has become prominent in stakeholder management discourse. To place the development of stakeholder management in Hong Kong in context, we examine how relationship management can shape stakeholder management and present two cases, as part of our ongoing research, to exemplify such an approach.

Literature review

Freeman (1984) contends that it is management's job to understand stakeholder behavior and to engage with them whether or not there is agreement on appropriateness of that behavior. Effective management of relationships with stakeholders is therefore an important managerial activity (Lim *et al.*, 2005). Relationship management has emerged as a sustainable approach for the construction industry in terms of people, environment and economics, and has the potential for satisfying client and stakeholder interests. This arises in part from the realization that the construction project can no longer be viewed as an isolated undertaking to satisfy the objectives of a small group of financing or sponsoring organizations, but must be viewed holistically as part of the social, economic and political structure within which it exists (Palmer and McGeorge, 1997). Managing the soft side of projects, such as the public image of major civil engineering projects, has thus proven to be as important as managing their physical creation (Lemley, 1996).

A relationship management approach demands a realization of the broadening of the boundary of the project organization where project managers are required to lead coalitions and coordinate interests which coincide, while resolving conflicts among non-aligned interests. It is essential, therefore, that senior managers do not view the large networks of stakeholders as constraints to the maximization of the organization's objectives, but must adopt the dominant managerial metaphor of negotiation (Freeman, 1984). Project procurement and financing arrangements have resulted in a shift from the singular client to plural client set-ups. An outcome of the increasing size and complexity of projects is that single construction organizations no longer have the capacity, resources and technical know-how to successfully implement such projects single-handedly. Joint ventures have therefore experienced a surge in places such as Hong Kong. In an ongoing infrastructure project for example, a combination of the above factors has resulted in a project organization set-up comprising over 20 primary stakeholder organizations including a plural client (four different departments fully involved), main contractor (joint venture of two organizations), consultant (resident site staff and the engineer), adjoining project 1 team (plural client – four departments, JV main contractor – four organizations and consultant) and adjoining project 2 team (plural client – four departments, JV main contractor – four organizations and consultant). These scenarios compound the difficulties in stakeholder management, as will be demonstrated later in case studies, making the need to employ relationship management principles in stakeholder engagement and management an imperative.

Over the past decade there has been an increasing emphasis on the use of relationship management in the administration of construction projects worldwide. This emphasis is manifest in the proliferation of partnering arrangements, public–private partnerships and alliances. Such approaches have met with varying degrees of success in different jurisdictions and this is in part due to the manner in which they have been implemented (see, for example, Cheung *et al.*, 2005; Lau and Rowlinson, 2005; Rowlinson *et al.*, 2006). Researchers have identified what they believe to be critical success factors for successful relationship management. Even so, there has been no consistent evidence on the efficacy of these factors. In particular, the influence of culture, be it sentient, organizational or national, has emerged as a strong moderating factor in the success of relationship management approaches.

Walker *et al.* (2008) have pointed out that the fundamental principles behind relationship management – trust building, commitment and innovation – are the same as those necessary for the implementation of successful stakeholder management. They illustrate this using an expectancy model and explain how trust is built up in phases and how both trust and distrust result in different management styles being used in dealing with relationships (Walker *et al.*, 2008: 79–80). The outcome of this process is the development of commitment, in this instance throughout the project team including stakeholders.

When dealing with the issue of relationship management it is, however, apparent from the literature that the choice of contract strategy has a significant impact on the effectiveness of the relationship management process (Cheung *et al.*, 2005; Rowlinson *et al.*, 2006; Walker and Rowlinson, 2008). Hence, when dealing with stakeholder management the same should apply. The case studies discussed below provide examples whereby the choice of an appropriate contract strategy facilitates the stakeholder management process. Indeed, one of the case studies indicates some serious shortcomings in the relationship management process, and similarly stakeholder management, because of an attempt to bolt on the partnering approach to a traditional design-bid-build contract.

In analyzing the nature of working relationships and management attitudes in an alliance project in Australia, Lingard *et al.* (2007) revealed the positive effects of a full-blown alliance on various aspects of individual and project performance. By reducing the working week to five days the work–life balance of employees was improved dramatically. As a consequence, a much more open and blame-free atmosphere developed amongst the participants in the project team than is usual on a construction project. This led to a more innovative approach to all aspects of work, enabling the project team to embrace the aspirations of all stakeholders in a positive manner. The conclusions that can be drawn from such a situation are that an appropriate contract strategy melded with a positive relationship management approach enables stakeholder management to take place in an atmosphere which is receptive and can find positive outcomes from divergent interests. Again, this finding is illustrated in the case studies.

While a relationship management approach as advocated above for stakeholder management has clear benefits, emerging empirical evidence in projects shows a range of response strategies being employed to engage and manage stakeholders. These generally range from proactive strategies consistent with relationship management principles to more passive strategies consistent with minimalistic interventions. In a recent study of stakeholder response strategies in four global projects, Aaltonen and Sivonen (2009) drew on the work of Oliver (1991) on organizational responses to institutional pressures to show five response strategies employed by focal organizations to manage stakeholders. The strategies include dismissal, avoidance, compromise, adaptation and influence. Dismissal strategies ignore stakeholder demands while pursuing project goals as defined by the focal organization. Avoidance strategies attempt to guard and shield the organization from stakeholder demands while deliberately transferring responsibility for responding to such demands to other organizations. A compromising strategy relies on negotiation and dialog to reconcile stakeholder demands with project goals and objectives. Under adaptation, the tendency is to yield to stakeholder demands leading to adjustment in project objectives and deliverables. Influence strategies however shape proactively the demands and values of stakeholders by actively sharing information and building relationships with stakeholders. Clearly, a relationship management approach to stakeholder management as discussed above aligns with compromise, adaptation and influence strategies. In the following discussion of two construction projects, traces of these strategies can be discerned.

Research project

Case study: Project Alpha

The case project is an integral part of a 7.6 km long major highway infrastructure undertaking. The works in Project Alpha comprise the construction of a 1.1 km elevated viaduct, dual three-lane carriageway (average 65 m above ground) to connect a tunnel (under construction) on one end and a cable-stayed bridge (under construction) at the other end. The project site is reclaimed land (to be handed over in phases), surrounded by industrial facilities, container terminals and an educational institution. The contract is a re-measurement type with a price fluctuation clause and awarded for an initial contract period of 40 months at an initial contract sum of HK$1,012 million. The project is delivered under a traditional design-bid-build approach in which the client engages the services of an engineering consultant to design, administer the contract and supervise the works undertaken by the contractor.

The particular features of this project, especially its size, location (vertically and horizontally) and technical complexity, brought together myriad stakeholders, whose interests needed to be aligned at various phases to deliver the project successfully. Five incidents, involving critical and contentious issues during the construction phase of the project, are used to illustrate how the stakeholders surrounding each incident were identified, managed or mismanaged individually and collectively in resolving the various issues. The impact of the procurement arrangement on the configuration of project stakeholders and the implications for their management are also discussed.

Incident analysis

Table 16.1 summarizes the key features of each incident: the stakeholders, stakeholder interests, consequences of not managing the interests, characterization of response strategy and manifestation of the response strategies. The nature of the incidents and their management are briefly discussed below.

Change in interface arrangement. The contractor proposed to change the interface arrangement regarding the positioning and maneuvering of the launching girder on the deck of an adjoining bridge project (under construction) from that proposed in its Technical Proposal at the bid stage and which was subsequently built into the contract. From an overall project perspective, the new proposal had implications for progress and risks, especially the achievement of the project's key dates. The contractor however considered the change necessary to make the launching operation simpler and safer. The stakeholders in this incident, whose input and buy-in was required, are summarized in the upper part of Table 16.1. The critical and contentious issues were:

1   structural stability of the bridge deck;
2   partial removal of temporary supports to the bridge deck;
3   achievement of key dates in jeopardy;
4   responsibility for risk and liability for any unforeseen circumstances; and
5   associated cost and time liability needing to be established.

The first two technical issues were easier to resolve with the bridge contractor, while the last three contractual issues were most problematic

Table 16.1 Stakeholder management initiatives and outcomes in Project Alpha

Stakeholder	Stakeholder interests	Consequences of not managing interests	Response strategy	Manifestations
	Interface arrangement			
Viaduct contractor	Safer work environment; Simpler site operations	Escalation of risks, non-achievement of key dates	Influence, compromise, adaptation	Buy-in of key stakeholders; formal and informal engagement; interface meetings; 'ping-pong' correspondence; presentations; mock demonstrations
Bridge contractor	Structural stability of bridge	Risk and liability		
Client	Limit liability and claims; Structural stability of bridge	Blame/reprimand from superiors; Escalation of risk		
Engineer's Representatives (viaduct and bridge)	Projecting client's interests; Enforcement of contract	Loss of client's trust		
The Engineers (viaduct and bridge)	Projecting client's interests; Enforcement of contract	Loss of client's trust		
ICE	Neutral assessment	Neutral		
Project Board of Directors (viaduct and bridge)	Safer and simpler site operations	Passive observer		
	Temporary Traffic Arrangement (TTA)			
Viaduct contractor	Non-compliance, least inconvenience to road users	Inconvenience to road users; loss of reputation of key project participants; public complaints	Influence, Compromise, Adaptation	Management of public expectations; three daily joint inspections; feedback from road users; complaint walk; Government's central complaints unit (1823 Citizens Easy Link (CEL))
Road users (general public)	Least inconvenience			
Client	Reduction in non-compliance, least inconvenience to road users			
Client's audit team	Enforcement of TTA			
Engineer's Representatives	Reduction in non-compliance, least inconvenience to road users			
TMLG	Faster resolution of TTA issues			

Stakeholder	Interest / concern	Risk / impact	Strategy	Actions
Community planting exercise				
Client	Community involvement; PR, promotion sense of ownership, public enthusiasm	Public agitation; negative publicity	Influence, compromise, adaptation	Invitation to participate; community out-reach; on-site community planting
Contractors	Liability and safety issues; insurance; composition of volunteers	Lack of commitment		
Engineer's Representative	*Projecting client's interests*; enforcement of contract	Loss of client's trust		
Public (school children)	Participation			
Construction Noise Permit (CNP)				
Contractor	24-hour cycle; constant supply of segments; storage area	Delays to works	Influence, compromise, adaptation	Mitigation measures; meetings; Government's central complaints unit (1823 Citizens Easy Link (CEL))
Client	Noise level; public complaints	Delays to works; public complaints		
School (Hall of residence)	Noise level	Inconvenience; public complaints		
EPD	Enforcement of noise regulation			
Miscast segments				
Contractor (pre-cast subcontractor)	Significant and unrecoverable delay and loss of resources	Delays to works Waste of resources	Influence, compromise, adaptation	Review of precast procedures; strengthening supervision; and mitigation measures
Client department/units (maintenance and audit)	Build as designed, easy maintenance	Maintenance difficulties		
Engineer's Representative/The Engineer	Enforcement of contract	Damaged reputation		'Ping-pong' correspondence
ICE	Neutral assessment	Neutral		

due to entrenched positions. Attempts to obtain buy-in of all parties included presentations and mock demonstrations, meetings and 'ping-pong' correspondence to resolve differences. The client required the contractors to waive their rights to claim time and associated costs which they declined. After six months of negotiations, the contractor was forced to revert to the original proposal. Ironically, the segment launching operation itself actually took less than three weeks to complete after reverting to the original plan. It is interesting that the spirit of the non-contractual partnering that was in place on the project and continuously reinforced through various workshops could not help. Indeed, an attempt to use the partnering process to resolve this issue was met with silence from all parties, reinforcing the skeptics' belief that many parties who sign up to such non-contractual partnering arrangements have little commitment to working in a true partnership.

While there appear to have been genuine efforts by the contractor (maybe because of standing to benefit most if the proposal was approved) to engage and obtain buy-in through response strategies, which can be characterized as involving influence, compromise and adaptation, it is doubtful whether any alternative mode of engaging (especially the client) could have yielded a different outcome. Public project settings are particularly replete with risk averse and fear of blame attitudes. This, rather than the means of engagement of the parties, may be why a proposal such as this was predisposed to failure.

Temporary Traffic Arrangement (TTA). To facilitate the works and safeguard the public, it was necessary during the project to temporarily divert traffic passing through the site. These changes to the normal movement of traffic are handled under a Temporary Traffic Arrangement (TTA). The key players and issues are shown in the middle of Table 16.1. The key stakeholder in the TTAs was the Traffic Management Liaison Group (TMLG), whose decision supersedes the contract provisions regarding the TTAs. The key players in the TMLG were the police and Transport Department, with the other members tending to go with whatever these two decided.

The client's audit team continually issued 'non-compliances' (NCs) for various breaches and the client's project team called on the Engineer's Representative (ER) to step up inspections to forestall any future breaches. The ER together with the contractor then instituted various measures to prevent contraventions of the TTA arrangements in the form of three daily joint-inspections – in the morning, afternoon and early evening. This was augmented with management of public expectations. Several initiatives were also in place in this regard:

- advance notice to client and concerned members of the public on TTAs;
- feedback from the public on TTA implementation; and

- 'complaint walk' where the client goes on site to walk through, with the ER and contractor, mitigation measures in response to complaints from the public.

These measures were successful in reducing the NCs to zero for the following months. TTAs are an important feature in roadwork projects and are considered one of the most challenging tasks on most road projects (Chan, 2003). Indeed, the project team, especially the contractor, is keen on ensuring that inconvenience to the public is reduced as much as possible by engaging all stakeholders for successful implementation of all TTAs.

Community planting exercise. Under a directive on 'community involvement in greening works', all capital works contracts with an estimated value of the landscape works in excess of HK$3 million should involve consultations with the respective district councils in regard to greening works prior to bid. It is a condition that the community be invited to participate in the planting works near to or after the completion of the project. Since the value of the landscape works on the project was less than HK$3 million, the adjoining bridge project (whose value for landscape was also less than HK$3 million) was invited to join the community planting exercise. Thus, both the contractor and consultant confirmed that the community planting exercise was not part of the original contract, but a public relations exercise by the client. Nonetheless, the ER was quite supportive.

The key participants for the community planting project were pupils from two selected primary schools in the neighborhood and some district council members. The contractor had some concerns, however, about the composition of the volunteers for the planting exercise and expressed reservations:

> there is some hidden risk in this, because for us at the moment, this is still a construction site; so under the law anybody who comes into the site will require a green card. If he is a worker, he needs to have a registration card ... the kids who will be doing the planting, they are actually doing *[the contractor's]* work. Technically they are doing our permanent works because they are planting the area where *[the contractor]* is supposed to plant, so they don't have green cards, they don't have workers' registration cards and they are all under age.
>
> (Contractor's representative)

Taken together, however, the community planting exercise appears to have been well received by the volunteers and attracted public enthusiasm. This can be attributed to the fact that it presents them with the opportunity to get closer to projects than they normally would, and in the process learn more about how taxpayers' money is being spent. Government and community representatives are also keen to show up at such exercises as it gives them the opportunity to engage closely and interact with their constituents.

Construction noise permit (CNP). Following a proposal to change from the use of two launching girders to one launching girder and a crawler crane, the contractor further proposed a 24-hour cycle for the erection of the viaduct segments in order to achieve an equivalent productivity level. The continuous supply of precast segments to the launching girder beyond 11pm in order to ensure that a 24-hour working cycle was achieved proved problematic, because the proposed storage area for the precast segments was directly beneath a student hall of residence and the carrier that supplied the segments to this area generated noise above the acceptable Environmental Protection Department's (EPD) limits. The key players and their interests are shown in the middle of Table 16.1.

To mitigate the situation, several measures exemplary of influence, compromise and adaptation were employed:

• modifications to the segment carrier using a noise enclosure;
• trial with measured noise levels recorded and presented to the EPD;
• closure of windows in the hall facing the site at all times; and
• replacement of old air-conditioners with much quieter new units.

The client played a key role in facilitating the approval process as testified by the contractor:

> [*the client*] was involved in some of the discussions, so everyone was involved trying to satisfy EPD, even [*the client*] went with [*the contractor*] to discuss with EPD, about what could be done, what is acceptable to [*EPD*] in terms of noise level from the point of view of EPD for it to issue a permit.
>
> (Contractor's representative)

Miscast segments. An estimated 67 precast viaduct segments were miscast by the precast subcontractor due to wrong setting-out information and resulted in the incorporation of cross-falls in the wrong direction. In view of the significant and unrecoverable delay to the work that this error could cause, there was the urgent need to review the procedures relating to the production of the precast segments in the yard in mainland China by strengthening supervision – see bottom of Table 16.1 for key players and interests in this incident.

Since some of the miscast segments were already erected, the key issues were to mitigate delays and consequences of the errors in the segments erected in terms of the alignment of the finished road surface. Given the implications of lost production time on progress of the works, the contractor further proposed incorporating as many of the miscast segments as possible into the works since the errors had no implications for the structural capacity of the viaduct. In line with this proposal, a full report on the segment errors was prepared and submitted to the ER so that the feasibility of further incorporating as many of the miscast segments (without rectification) into the works could be evaluated.

There was close collaboration among all parties to resolve this issue as soon as possible and the client's role was especially crucial. It is clear that the consequence of the miscast error for all stakeholders was an incentive to work together for a fruitful resolution of the problem. This demonstrates the power of joint interest or joint risk in motivating stakeholders to work for the common good of the project. Yet, the inability to agree on how to dispose of the remaining precast segments showed how lack of alignment of interests forestalls consensus building.

Impact of procurement arrangement

As noted earlier, the project was procured under a traditional design-bid-build approach. It is apparent from the discussion so far that the arms-length mindset associated with this approach contributed to how some of the incidents played out. It is commendable, however, that interface arrangements were built into the contract. This approach clearly defined the interdependence between the two projects from the outset as an issue to be managed. Nonetheless, the interface arrangement appears to have been structured without consideration for the uncertainties that can arise in a project of this size and complexity. The situation was further exacerbated by the inflexibility of the various parties. Ironically, there was a non-contractual partnering arrangement in place, in which the parties promised to work in partnership. Yet, when it mattered most all the stakeholders held on to their contractual rights.

The structuring of the project organization also had implications for a number of stakeholders on any issue and thus their management. First, the client organization was pluralistic. On many issues three or more different departments of the client organization needed to be satisfied and this became even more problematic when they disagreed. The fact that the contractors on the two adjoining projects were joint ventures also had implications for engaging them. In this case, the board of directors of the JVs appeared to have played only a passive role, as most of issues were considered site matters, which were within the domain of the site teams. Some contractual provisions also had implications for the number of stakeholders who needed to be engaged; for example, the Engineer's Representative as a separate entity from the Engineer and the use of an Independent Checking Engineer (ICE), whose role was to check all contractor designs and the TMLG.

Stakeholder management outcomes

Five incidents have been analyzed above to show how stakeholder management on an infrastructure project manifested (see Table 16.1 for full summary). In the management of both internal and external stakeholders, it was clear that when the stake for all stakeholders on the issue of contention was high there was a tendency to reach an agreement easily. Culture-specific dynamics also manifested in the positions that different stakeholders took on issues and there was a general tendency for 'rule following' or adherence strictly to the contract. This may be attributable to the fear of blame culture pervasive in public project settings and the conflict avoiding view inherent in the Confucian value system.

Notwithstanding the good intentions of proponents, the incidents also indicate that it might sometimes be impossible to gain buy-in from stakeholders no matter how hard parties try to engage. Buy-in appears particularly difficult when the issues are contractual in nature. The need for stakeholder management is also driven in some cases by government policy or contractual arrangements (e.g. TTAs, interface arrangement and community planting). While this may give parties the opportunity to strategize and implement more structured approaches to managing stakeholders, the incidents show that ad hoc approaches are dominant. Unlike the projects analyzed by Aaltonen and Sivonen (2009) however, the response strategies employed by the focal organizations were proactive, reflective of a desire to invoke relationship management strategies in managing stakeholders.

Taken together, this case study demonstrates an element of progress toward public engagement on projects in Hong Kong, an element that was unheard of a decade ago. Yet, the arms-length mindset, perpetuated by decades of use of the traditional procurement approach, is still prevalent. Indeed, when collaborative initiatives such as partnering are bolted on to the traditional procurement system there is little evidence of real partnership. Thus, a shift in culture, both in terms of the way stakeholders are engaged and projects procured, appears a viable option for project delivery in Hong Kong.

Case study: Project Beta

The project is being implemented at a time when there is increasing emphasis by the Hong Kong Special Administrative Region (HKSAR) government on sustainability and community development in public housing through the procurement and implementation of project processes. Four sustainability dimensions have been adopted by the government with a focus on balancing the economic, environmental and social concerns of all the stakeholders in the project. To achieve these goals, various initiatives are increasingly being embedded in the bidding and contracting procedures in the implementation of projects.

Project Beta is Phase 4 (of six phases) of public rented housing involving the construction of three 41-storey blocks, estimated to provide a total of 2,369 units of rental apartments. The value of the works is estimated at about HK$434 million and is contracted out for an initial period of 36 months. The works are procured broadly under a traditional design-bid-build approach using the Government of Hong Kong General Conditions of Contract for Building Works (1993 Edition). Special conditions of contract are incorporated to cater for six work packages contracted under a modified Guaranteed Maximum Price (GMP) arrangement which collectively makes up about 31 percent of the contract sum.

Stakeholder management

Several initiatives were implemented to engage stakeholders both internal and external to the project organization. Table 16.2 provides a general summary of the key issues, the stakeholders, stakeholder interests, consequences of not managing their interests, characterization of response strategy and manifestation of the response strategies. The first initiative targeted at internal stakeholders is the 'workers wage protection scheme'. The scheme is a direct response to workers' concerns on the protection of their wages in the event of default by the contractor or subcontractor. This scheme had several elements:

- on-demand bond in the contract which can be used to secure payment of wages for affected workers;
- a labor relations officer (LRO) employed on site to check, verify and monitor workers' wage records. The LRO also receives, acknowledges and records complaints and follows up complaints on site;
- subcontractors are required to pay their workers on time before applying to the main contractor for their monthly payment in conjunction with works done; and
- computerized wage monitoring system equipped with a sophisticated mechanism to track wage payment such that if late payment to the workers is encountered, the system issues a warning and the subcontractor's payment is delayed.

The main contractor of the project has also adopted other primary stakeholder management initiatives concerning mainly the on-site welfare provision for workers and staff, and human resource development for the site management team. The initiatives include:

- health promotion program that includes basic health check and health counseling for workers with health conditions (e.g. hypertension); cash prizes for high performing workers; heat stress preventive program in view of the high-temperature summer working periods; the provision of mobile mist-generating machines; installation of thermometers throughout the site; the provision of workers' quarters and laundry areas;

Table 16.2 Stakeholder management initiatives and outcomes in Project Beta

Stakeholder	Stakeholder interests	Consequences of not managing interests	Response strategy	Manifestations
Procurement specific initiatives				
Client	Embraced contractor expertise, improved buildability	Less buildable design	Adaptation	Design-and-build element in the GMP packages
Client	Cost certainty, risk reduction	Cost escalation	Adaptation	Introduce GMP scheme
Main contractor	Equitable cost and risk sharing	Cost escalation	Influence	Client administers pain-and-gain share scheme
Subcontractors	Enjoy the benefit of saving	Less motivated to suggest buildable design	Influence	Client administers pain-and-gain share scheme
Client, main contractor and subcontractors	Better disputes resolution	Cost escalation, delay, and negative relations among parties	Influence, Compromise	Client administers dispute resolution advisor system
Client, main contractor and subcontractors	More amicable working environment	Negative and adversarial working relationships	Influence	Developing team spirit through project partnering
General initiatives				
Workers	Prompt payment of wages	Low morale, work stoppage	Influence, Compromise	Wage protection scheme
Workers	Welfare and safe working environment	Low morale, lost productivity due to incident/ accident	Influence, Adaptation	Main contractor provides safe and comfortable working environment, health promotion

Stakeholder	Need/Interest	Negative consequence	Strategy	Action
Project team members	Self-improvement and promotion	Low morale and productivity	Influence, Compromise	Main contractor implements human resources development
Project team members	Familial working team	Low morale	Influence, Compromise	Main contractor's project manager promotes team cohesion
Client and main contractor	Organization and company image	Negative publicity	Influence, Compromise, Adaptation	Active engagement with community and public to improve communication and impression
Client and main contractor	To be recognized as socially responsible corporate entity	Bad corporate image	Influence	Active implementation of corporate social responsible activities
Community and public	Participation in the development of estate	More complaints	Influence, Compromise, Adaptation	Client and main contractor's engagement activities and communication sessions
Community and public	Less disruption of their living environment	More complaints	Influence, Compromise, Adaptation	Noise and dust reduction construction methods
Elderly residents at adjacent estate	Malfunction within-unit services repaired at low or no cost	n.a.	Influence	Main contractor free attendance to the units

- team members are encouraged and sent to attend various personal development courses that include management skills, technical skills and leadership;
- promotion of a familial atmosphere among the site team, e.g. coaching program, recognition and the active seeking and provision of opportunities for site staff to try new things within their capability;
- systematic recognition and promotion scheme (both financial and positional rewards). The results observed were the promotion of some site staff and the re-joining of some junior engineers after the completion of their industrial training with the main contractor.

The emerging outcomes of these initiatives are in line with studies conducted elsewhere that indicate the clan-type culture which emphasizes that people orientation is more conducive to successful project outcomes, albeit in the area of quality management (Thomas *et al.*, 2002). The management of secondary stakeholders, in particular, on the part of the client has seen a saliency in the client's proactiveness in engaging stakeholders. The client has built into planning and development processes a number of community engagement initiatives:

- a series of activities designed to instill a greater sense of belonging and participation of the community in the project, e.g. a competition for mural painting was organized in the community with the winning design being incorporated as a permanent mural feature for the estate;
- 'Action Seedling' to promote community participation in the project. Local residents and schoolchildren from nearby schools participated in planting seedlings and nursing the plants for the estate under construction; and
- extensive use of prefabricated building elements and hard paved site areas to reduce dust and noise.

In response to the client's push for active community engagement from the beginning, the notion of corporate social responsibility (CSR) gradually evolved throughout the main contractor's organization. As a result of increased awareness on the impact of its activities on the community, the contractor has been active in participating and responding to the client's drive for community engagement, at times going beyond the requirements of the client. Two incidents exemplify the contractor's active involvement in volunteer activities.

1 House improvement work during a festival to help elderly residents at the nearby estate, by dispatching two teams of personnel to help repair malfunctioning services within the apartments.
2 Construction related information provided to nearby residents in connection with prolonging construction activities beyond normal working hours.

Stakeholder management outcomes

Several implications can be drawn from the foregoing project stakeholder management initiatives in this project. As in Project Alpha above, stakeholder management responses in Project Beta tended to embrace a relationship management perspective. Even so, not every level within the organization exhibited this proactive attitude as elements of dismissal and avoidance surfaced at the lower levels as exemplified by some of the outcomes discussed below.

Passive reaction. There was passive reaction among the subcontractors and junior staff members to the initiation and implementation of stakeholder management. The situation was particularly evident in the management of secondary stakeholders. It appeared that the members of the lower echelon were adopting a minimalist approach. For these members, engaging with external stakeholders was not seen as contributing directly to their immediate work.

Lack of a structured approach to project stakeholder management. The preceding observation is symptomatic of the lack of a structured project stakeholder management system on the part of the main contractor. The deficiency is particularly acute with external stakeholder management. Despite considerable success in dealing with and tackling issues within the community, the main contractor admitted that the approach was one of trial and error. There was no deliberate or structured approach to identify external stakeholders, their impacts and the method of engaging them. While the efforts and achievement of the main contractor have to be commended, the situation reflects the somewhat parochial mentality of the construction fraternity in terms of external stakeholder management.

Contracting firms have traditionally adopted the attitude that construction operations are confined within the boundary of the site. Site operations are therefore a closed system. This view overlooks both the direct (e.g. dust and noise) and indirect impact (e.g. bad impression resulting from direct impact) on the community. In terms of engaging external stakeholders and mitigating the impacts construction activities cause, it is not in the interest of firms to do more than necessary as costs are incurred in extra efforts. Hence, shareholder management and interest still overrides the stakeholder paradigm. That is, the stakeholders' perspectives are not integrated into the project formulation processes despite the best intentions of both parties (cf. Cleland and Ireland, 2007).

No allowance for additional resources for stakeholder management. Despite the various external stakeholder management activities that had been carried out by the main contractor, there was no provision of additional resources for the main contractor under the contract. The reward from the client comes in the form of recognition. In addition, given its status as a pilot project the ensuing image issues and the high stakes involved, especially for the two primary stakeholders of the client and main contractor, the latter resorted to absorbing the extra costs

(Mahesh *et al.*, 2007). Yet, while the costs involved in carrying out those activities are not considered large, the lack of compensation from the client may lead to token efforts from the main contractor.

Engagement of specialist subcontractors from the client's nominated list. The subcontractors for two GMP packages were 'novated' from the client's nominated list, but because of the nature and element of design and build inherent in the packages, these subcontractors were engaged as domestic subcontractors. The arrangement is seen as a move to improve constructability, thereby achieving a cost-saving design. Although the arrangement helps ensure quality control to some extent for the client, it can reduce the main contractor's capacity to stay within the GMP (Haley and Shaw, 2002). In addition, the level of cooperation between these novated subcontractors and the main contractor needs extra attention and promotion. For this project, it was observed that the client's intervention was invoked in the initial stage of the project to bring the parties together. In the long run, however, a more appropriate arrangement needs to be implemented.

Discussion

It is apparent from the case studies above that tradition, custom and practice, politics and culture have a major influence on how stakeholder management is undertaken in the Hong Kong construction industry. Without a strong tradition of democracy it is not surprising that the move to draw the public, green groups and other parties into the development process has moved forward slowly; there is no evidence of resistance to change, rather an inertia grounded in the traditional values of society and the structure of government departments and institutions which puts a brake on change. This is not totally surprising: if one studies the position of Hong Kong on Hofstede's dimensions of culture it is obvious that nations such as the UK and the USA have a value infrastructure which is more open to stakeholder involvement and empowerment (see Figure 16.1). The Confucian values of harmony and conflict avoidance are often an opposing force to the drive for stakeholder empowerment.

There is evidence from the case studies that change is taking place and that the post-colonial administration is becoming more attuned to the legitimate demands of its stakeholders and a re-education process is taking place. It is apparent from the cases discussed above that focal organizations are shedding their dismissal and avoidance response strategies of the past and embracing proactive responses of adaptation, compromise and influence to manage stakeholders. This cannot be described as a cultural revolution, but a culture change is taking place. A move away from traditional procurement forms is now underway with the Hong Kong Housing Authority leading the way and the other Works Bureau departments commencing a range of experiments with more open procurement forms. Indeed, the incorporation of partnering type agreements into many projects

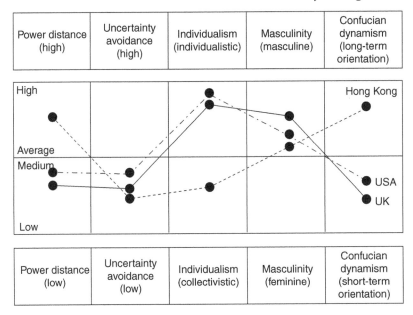

Power distance (high)	Uncertainty avoidance (high)	Individualism (individualistic)	Masculinity (masculine)	Confucian dynamism (long-term orientation)

Power distance (low)	Uncertainty avoidance (low)	Individualism (collectivistic)	Masculinity (feminine)	Confucian dynamism (short-term orientation)

Figure 16.1 Comparison of scores for Hong Kong, the USA and the UK on Hofstede's cultural dimensions.

has contributed to a change in culture and led to more open attitudes to cooperation and collaboration in construction projects (Anvuur, 2008). In line with this development there needs to be a recognition that performance measures have to be refocused to take into account medium- and long-term objectives in line with the arguments put forward by Walker *et al.* (2008). As Table 16.3 shows the stakeholder management strategies in both cases were driven by five main issues: procurement systems reform, improved collaboration, life cycle value consideration, community involvement and community benefits.

In recent years, employees and stakeholders have also become much more aware of the need for firms and government to show a commitment to corporate social responsibility (Rowlinson, 2008) and this has raised awareness in all sectors. Indeed, major infrastructure and real estate developers have taken on board stakeholder management as part of their corporate social responsibility commitment; time will tell whether this is a marketing fad or a genuine culture change in the industry. With the establishment of the Construction Industry Council in 2008 there is now an industry-wide body dedicated to improving performance in the real estate and construction industries. One of its first tasks has been to improve construction site safety and this has involved an attempt to engage workers, managers and directors in a framework that provides a basis for joint problem-solving and initiative development. Such approaches augur well for the future development of stakeholder management and empowerment.

Table 16.3 Comparative analysis of stakeholder management issues and strategies across cases

Issue	Case	Response strategy	Examples	Implementation
Procurement systems reformation	Beta	Influence, adaptation	Greater contractor participation, equitable sharing of costs and benefits	MGMP packages, dispute resolution system, pain-and-gain share scheme
	Alpha and Beta	Influence	Partnering	Non-contractual partnering; interface management
Improved collaboration	Alpha and Beta	Influence	Relationship management	Partnering, promote dialog sessions
Lifecycle value consideration	Beta	Adaptation	Lifecycle costing	Design with maintainability in mind
	Alpha	Influence, compromise	Emphasis on what is best for the project in the long run	Build with maintainability in mind (incorporation of miscast segments); Owner Controlled Insurance Programme (OCIP)
Community involvement	Beta	Influence	Proactive engagement, greater community participation	Community planting, mural wall design, dialogue sessions, volunteering information
	Alpha	Influence, adaptation, compromise	Buy-in of key stakeholders; formal and informal engagement; Management of public expectations; community out-reach	On-site community planting; Government's central complaints unit (1823 Citizens Easy Link (CEL))
Community benefits	Beta	Influence, compromise	Provision of direct and indirect benefits	Low dust and noise generating construction methods, free house improvement services, improved greenery around construction site
	Alpha	Influence, adaptation, compromise	Buy-in of key stakeholders; formal and informal engagement; Management of public expectations; community out-reach	On-site community planting

Conclusions

For further progress to be made in stakeholder management the Hong Kong real estate and construction industry needs to build on the current modest achievements as exemplified in the two cases above. This will require that the industry addresses several knotty issues that continue to inhibit progress in effective stakeholder engagement and relationship management. A good starting point is procurement reform. There is the need to allow for more innovative and collaborative approaches to the project development process. This should then be extended across all the phases of the project process so that a culture change can begin to take place throughout the industry where participants focus on cooperation and collaboration rather than defensive reactions.

As the two cases show, a focus on the real meaning of value in the project context rather than a decision-making process based on lowest initial costs is a much more promising path for the industry. Such an approach will reinforce the cooperative and collaborative agenda, allowing a focus on what is best for the project. At the front-end of project implementation then, a commitment to community involvement and a full implementation of the principles of corporate social responsibility in both public and private sectors will be required. This will also mean that organizations empower the teams and individuals they deploy at the project level and who interact at the organization interfaces so they can effectively engage each other and the external stakeholders of the projects.

17 Learning in demonstration projects for sustainable building

Barbara Rubino and Michael Edén

Introduction

As a key feature of sustainable construction, energy efficiency is both an objective and a problem to be solved within the project. As a problem, it is accompanied by many sub-problems and can easily become ill-defined. It can be seen as an innovation and as a technological change. Ideally, it would be better to understand both simultaneously, but that is difficult. At one extreme, energy efficiency is a necessary worldwide goal, the realization of which requires a global commitment toward sustainable development and building. At the same time, it requires numerous innovations (technological as much as conceptual and strategic) and change to be implemented inside local contexts. Both innovations and change have to penetrate the disciplinary organization of knowledge, inducing change toward sustainable building, while respecting and understanding the heterogeneous engineering that is needed.

The aim of this chapter is to provide a distillation of observations about the kind of learning going on in projects in terms of its elements and procedures. Investments in learning may be questions of providing more time for project teams, but also of encouraging actions of deliberate learning in project environments. The study is based on empirical observations made within the context of a funded research project.

Literature review

The nature of practice in projects is considered here to be central to the implementation of approaches to sustainable building. In this regard, there is need for new clarifying descriptions as to learning processes and the development of technology within defined limits. The transition to more sustainable building practices appears as 'heterogeneous engineering', using a term introduced by John Law (Bijker and Law, 1992). The term can be conveniently assumed to give us the impulse to look for all factors that are needed for the successful implementation of an idea. Heterogeneous engineering seeks to associate entities that range from people and skills, to artifacts and natural phenomena. Taking a step aside from other, more usual approaches to analysis may offer up new and useful understandings. Law maintains that for a real innovative technical change to succeed, a non-linear combination of interacting casualties is needed (Bijker, 1987). 'From this point of view, sustainable building will succeed in increasing its energy efficiency when it becomes *a network of artifacts and skills* converting small quantities of energy into allies in the struggle to master the power of climate' (Rubino, 2006a). Law suggests that heterogeneous engineering needs three stages:

1   the *process of shaping technologies*, as applications and inventions, and *scientific knowledge*;
2   *social engineering* which constructs a network of practices associated with developed tools and components, broadening the field of contextual applications, converting *esoteric scientific knowledge* into a widely applicable practice and identifying the *weakest link* in the attempt to create a stable network of elements; and
3   definition of a *point of return*, meaning that all decision-making during the process may not be possible without a *scale of reference – technological testing* implies the construction of a background against which to measure success.

From this perspective, it may be assumed that the construction industry is currently challenged by a societal push toward a technological change that has become political and economic. Energy efficiency seems to be the core question in this change.

Energy efficiency in buildings is hindered by knowledge barriers (Nässen and Holmberg, 2005; Femenías, 2004). Knowledge about specific issues is regrettably lacking in some of the parties involved (e.g. clients, architects and builders). The assumption is that it makes use of experience and depends on feedback mechanisms. The need to transfer experience in different directions within and between organizations in the industry has been identified by some scholars (e.g. Sprei, 2007), while others point to the management of organizational capabilities in order to diffuse new knowledge and competence inside organizations (Zollo and Winter, 2002).

An impediment that is often highlighted, mostly by practitioners, concerns the many different measurements used for different aspects of energy efficiency in buildings. Normally, energy use is measured in terms of KWh/m^2yr for heating, hot water or the two together for household electricity and total energy consumed. Yet, energy efficiency also concerns the insulating properties (U-value) of materials and/or components of whole parts of the building. Energy efficiency is finally also affected by life cycle cost analyses of various materials, components and technical assemblies.

Scientific knowledge has been developing over decades, even if the particular knowledge domain of sustainability is so new that all agendas have to be revamped. Still, the framework of studies about sustainability belongs, by nature of its origin, to the scientific domain and constitutes esoteric scientific knowledge that is difficult to translate into widely applicable practice. The processes of shaping technologies and scientific knowledge do not seem to develop hand-in-hand. Technologies in construction appear to be shaped as a part of the process of social engineering a change into contextual applications and networks of practices. A huge amount of practitioner research within project teams is actually shaping technologies in construction, confining scientific knowledge to the theoretical formulations of rules, problems and solutions led by analytical rationality. Scientific knowledge developed in academic environments, on the other hand, has never really accepted the fragmented, not easily accessible, knowledge environment of practitioners, who complain about their resistance to change and innovations as well as about the diffused, non-rational approach used in problem solving and decision-making.

The past decade has seen an increased interest in the introduction of social science theories into the field of building research. As a result, projects are now studied as complex social settings, as learning environments and as communities of practices where specific collaboration is taking place (Bresnen *et al.*, 2005). A cognitive approach is widespread, essentially in terms of its supposition about the participation of individuals with their beliefs, assumptions, history of experience etc. In looking at projects as time-dependent activities, which produce learning more than knowledge, the focus has discarded many hypotheses or definitions of knowledge as codified, tacit or articulated, giving power to one or the other interpretation (Wenger, 1998; Koskinen *et al.*, 2003; Prencipe and Tell, 2001).

Project-based approach for research and practice

Aims and objectives

The overall aims of the research are two-fold:

1   to understand the role played by demonstration projects in the implementation of sustainable building practices; and
2   to shape a multi-actor arena for the people-to-people enactment of the different hypotheses, new models and solutions proposed in these projects.

Demonstration projects are investigated as an example of social engineering within ongoing technological change toward sustainable building, where theoretical context-independent knowledge is translated into context-dependent practice.

Demonstration projects are considered as the contexts for change, since a large amount of learning is generated and can be monitored in the interactions between heterogeneous actors/agents, skills and technological components, with the opportunity to connect them as a whole in responding to a need. Results are dependent on the initial definition of goals (mostly centered on energy efficiency, both quantitative and qualitative) and on the achievement of outcomes (always quantitative) which must be scientifically tested in order to measure the success of the project overall.

The field of investigation is an example of local construction reality and comprises six projects which all define themselves as wanting to provide examples for the future. They are evolving at the present and are concrete. The agents, actively participating in the research investigation, are working as engineers, architects, clients and developers and have within the framework of the project emerged as individuals, as representatives of firms and companies, and as members of project teams representing different ways of working, decision-making, problem solving and learning. The projects define the context as one of parallel distributed processes. These could reveal patterns that might help in finding answers to our research questions.

Multi-mode approach

The approach has been to gather a network of practitioners, who have carried out demonstration projects, in order to exchange experience and, hopefully, to establish more efficient processes for internal learning as well as for external communication. A group of 'change agencies' was tied to the network early on. They represent stakeholder organizations that, in different ways, conduct information or educational activities. These are meant to collect the findings from the projects and transform them into different kinds of learning material that can be used to initiate either the mainstreaming of sustainable building or new demonstration projects.

The research kick-off meeting provided the guidelines for the development of a qualitative multi-mode method comprising:

1 semi-structured interviews with the practitioners and the clients;
2 focus-group meetings, organized and held on specific issues;
3 feedback seminars with end-users of finished demonstration projects or experts; and
4 follow-up on project meetings for three cases, of which one has been reported (Rubino, 2006b).

Together they shape what we call a 'multi-actor arena'. All meetings have been monitored and followed up with notes, and a draft has been sent to all participants for review. The height of learning has been reached in workshops, where feedback to projects was provided by users and real estate owners. Interest and participation were achieved at a conference at the mid-point when the demonstration projects were presented to a broader public. Engineers, architects, clients and project managers presented aspects of their work in terms of the problems faced, the tools developed, ideas and provisional results.

Research methodology

The research project has been structured around features that categorize it as action research. This has initially been due to its objective of fostering change toward sustainable development. At the same time it has many of the typical characteristics discussed about action research when trying to give definitions of it; for instance 'action research is not a method or a procedure for research, but a series of commitments to observe and "problematize" through practice, a series of principles for conducting social enquiry' (Malterud, 1995).

> The research needed for social practice can best be characterized as research for social management or social engineering. It is a type of action-research, a comparative research on the conditions and effects of various forms of social action, and research leading to social action. Research that produces nothing but books will not suffice.
>
> (Lewin, 1946, reproduced in Lewin, 1948: 202–3).

Lewin's original approach involves a spiral of steps, 'each of which is composed of a circle of planning, action and fact-finding about the result of the action'. The basic cycle involves the features adapted from Malterud (1995) (Figure 17.1).

The action research spiral was never adopted as a leading method; hence, it can be useful as an explanation of what has been happening during the research. It is a *processual* study and has resulted in overwhelming

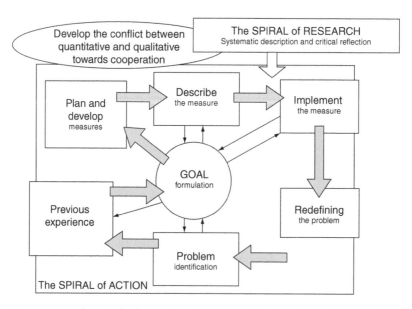

Figure 17.1 The spiral of steps in action research (adapted from Malterud, 1995).

empirical material. Working with empirical data can become a little like treading on well-known paths making us blind to unknown pathways, unless we actively search for them.

Follow-up study

It was a follow-up study that led to the questioning of *unknown pathways*. The follow-up was considered likely to confirm the initial hypotheses that a clear vision, best practice attitude, well-defined problems and goals, and transparency of results were the *rational recipe* for energy efficiency and for the implementation of environmental goals in buildings. The project looked like the model enterprise – the single project in which all preconditions were satisfied and in which all regulations on environmental and energy efficiency performances had been fulfilled. The model of deductive, analytical rationality was expected to be satisfied in this kind of organization. On the surface and inside a posteriori reconstruction, the model appears to apply and is theoretically supposed to apply in construction practice, but was strangely absent in the in-depth details of the case. What was instead present within the project was an enormous amount of practice-related research going on at every stage, together with the human ability to cope with complicated situations and the enormous incentive given by simply formulating 'we are demonstrative!'. The actors involved did not sleep at night, exalted by the challenge of solving a problem.

The study became a critical case (Flyvbjerg, 2001) as its conclusions became a matter of: 'If they find it difficult then surely all others do too!' They were supposed to show the *how*, the right way; knowledge and technology were at their disposal. Moreover, it was likely that the same problems would exist in other projects that were less careful with rules and goal setting.

A series of random events showed up under closer examination of the process. Decision-making was never linear. A number of hypotheses were tried and discarded, the formulation of problems changed continuously, the lack of objects to refer to was frustrating and an elephant was going to be built with an ant as the model. Furthermore, important decisions seemed never to be made within the project team. The team regularly presented integrated hypotheses, the messages in which were filtered by other 'groups', such as the leadership group, the 'group for the silhouette of the town' and the group for the total environmental management. The use of energy performance components and solar panels, general layout, car and cycle parking solutions were determined in separate rooms in terms of their performance as cultural signs, political statements and end-user friendly solutions.

The individual learning which was going on within the project was strictly concerned with the process of testing hypotheses and rejecting them. Feedback from others in the group seemed to be crucial and recursive; each project meeting was an outcome and a point of departure on the path toward a solution. The way from the formulation of a problem to its solution is difficult. Monitoring a project from within offers the chance to observe agents engaged in the act of taking decisions. The follow-up study produced a lot of empirical evidence about the messy process we try to describe. Although it happens, it does not make sense or have any rationale. The empirical results combined with the analytical model elaborated by Arthur (1994, 2000, 2007) gave the basis for the following tentative description model.

Inside the black box

A project is what is to be found between the formulation of a problem and its solution. It is a social setting that unfolds in time, following a structure which is not necessarily its formal organization. Projects are self-organizing settings in the direction of a reification of the participative work of agents with heterogeneous origins. Moreover, projects are time-dependent processes. As the project for a new building unfolds, different rationalities interact with increasing complicatedness. Agents organize in order to cope with problems, which are always transformed into ill-defined problems, with bounded rationality.

Rationality and the definition of projects

There is a diffuse tendency among both practitioners and academics to call for well-defined problems and rational, deductive analysis. Changes in the environment of project work are recurrent. Objective, well-defined, shared assumptions are expected to characterize the process. A belief in human rationality – perfect, logical and deductive – is rooted in an aspiration to control the outcomes of projects.

If one sets up a problem and assumes rationality in decision-making, a well-defined solution is expected to follow. The act of building is simple from this perspective: from the problem flows the solution. How agents get from problem to solution is often considered a 'black box' and whether, indeed, agents can arrive at the solution cannot be guaranteed unless we look into this box. If we open it, building becomes difficult. Complications arise. 'Rationality, so useful in generating solutions to theoretical problems, demands much human behavior, more than it can usually deliver'. Rationality breaks down under complications (Arthur, 1994, 2000). This used to happen under two conditions. First, beyond a certain complicatedness the human logical apparatus ceases to cope and rationality is bounded. Second, in interactive, complicated situations agents cannot rely upon each other to behave under perfect rationality and so they are forced to guess their behavior. Objective, well-defined, shared assumptions cease to apply. 'In turn, rational, deductive reasoning – deriving a conclusion by perfect logical processes from well-defined premises – cannot apply. The problem becomes ill-defined' (Arthur, 1994).

Opening the box – a cognitive approach

If a deductive approach and perfect rationality have limited functions then the real problem is what people place instead of rationality, not whether or not perfect rationality works. From behavioral psychology, we learn that in situations that are complicated or ill-defined, humans use characteristic and predictable methods of reasoning. These methods are not deductive, but inductive. This assumption makes a great difference and as Arthur (1994) suggests 'it makes excellent sense as an intellectual process; and it is not hard to model'. Individuals together find optimal solutions to complex questions.

Professional decision-makers do not back off from a problem because it is difficult or unspecified. Solutions may stop matching reality and stop existing as such. When problems are too complicated to afford solutions or when they are not well specified, agents do not face a problem but a situation. They must deal with that situation; they frame the problem and that framing is, in many ways, the most important part of the decision process. Cognition more than knowledge or learning is at play here. Mind, associations and meaning are central questions of a cognitive approach, but not at the level of this inquiry. Hence, one element is relevant for understanding what happens inside the project: people do not only think deductively, people think associatively and use inductive reasoning.

Induction, hypothesis and learning

How do agents cope with situations that are complicated or ill-defined? Psychologists say that they see or recognize or match patterns. Mostly, humans look for patterns and simplify problems by using them to construct *temporary internal models*, *hypotheses* or *schemata*. Agents shape localized deductions based on current hypotheses and act on them. Feedback from the team comes in and may strengthen or weaken their beliefs in their current hypotheses, discarding some when they cease to perform and replacing them as needed with new ones. In other words, where they cannot fully reason or lack full definition of the problem, they use simple models to fill the gaps in their understanding. This kind of behavior is inductive.

As the project unfolds, agents hold on to hypotheses or mental models that prove plausible or toss them aside if not, generating new ones to put in their place. In other words they use a sequence of *pattern recognition > hypotheses formation > deduction using currently-held hypotheses > replacement of hypotheses* as needed. This behavior enables agents to deal with complications when constructing plausible, simpler models that they *can* cope with. It also enables them to deal with any lack of definition. When they have insufficient definition, their working models fill the gap. Arthur (2007) claims that, in fact, this is the way science itself operates and progresses. This research project also follows this behavior pattern.

It is possible to model induction. All building problems are problems that play out over time, with a collection of heterogeneous agents. We can assume that agents form mental models, hypotheses or subjective beliefs. These beliefs may come in several forms: mathematical expressions that can be used to describe or predict some variable or action; statistical hypothesis; or prediction rules. They are normally subjective, which means that they differ among agents. Moreover, an agent may have one in mind at a time or a number of them at the same time. Each agent normally keeps track of the performance of his/her private collection of these 'belief models'. As the project unfolds and choices are made, he/she acts on the currently most credible, most profitable one. He/she keeps the others at the back of her/his mind. Expert professionals generally hold many hypotheses in mind and act on the most plausible. Problems may arise when only a few or no hypotheses are on hand. This procedure makes action possible and when action is taken 'an aggregate picture' is updated as a possible solution and agents update the track records of their hypotheses. Each project meeting is a point of transition. The process is path dependent.

Learning takes place inside this system. Agents learn which of their hypotheses work and discard poorly performing hypotheses from time to time. In their place, they put newly generated ideas which are applied in the current project or not – they may be applied in the next project. The currently most plausible hypotheses and belief models are clung on to until they no longer function well and then they are dropped in favor of better ones. A belief model is not 'clung on to because it is *correct* – there is no way to know this – but rather because it has worked in the past and must accumulate a record of failures before it is worth discarding' (Arthur, 1994). There is evidence for a general, constant, slow turnover of hypotheses acted upon. Arthur calls them a system of temporarily fulfilled expectations, and he means that the beliefs, hypotheses or models are only temporarily fulfilled and never perfectly fulfilled, and that this fact opens the way to different beliefs or hypotheses when they cease to be fulfilled. Where do these hypotheses or mental models come from? They are generated behaviorally in cross-fertilization between cognition, object representation and pattern recognition. Agents are somehow endowed with 'focal models' (Arthur, 1994), i.e. patterns or hypotheses, which are obvious, simple and easily dealt with mentally. A bank of them can be generated and distributed; but they have to be obvious, simple and easily dealt with and are not examples to be applied, but instead something to focus upon. They are not solutions, rather tools with which to work. Focal models may become worth discarding. As problems become ill-defined so do the expected solutions.

Technology at work in projects

Technology is not necessarily developing parallel to knowledge and the way it works may also depend on interacting behavioral processes of inductive reasoning. Technology is a means to fulfilling a human purpose (Arthur, 2007). The purpose can be explicit, well-defined or not. Technology is also a body of practices and components: the totality of the means employed by people to provide themselves with the objects of material culture. Hence, as a means to fulfill a purpose, 'a technology may be a process or a method or a device' (Arthur, 2007). It is normally put together or combined from component parts or assemblies, sub-systems and sub-technologies.

Technology always proceeds 'from some central idea or concept – the *method of the thing*' (Arthur, 2007). This is the base principle or base concept of the technology and does not need to be simple. 'Passive house' is a base principle for building, from which the choice of technology proceeds. A principle is an idea, a concept, which the agents within a project agree on, adapt to or shape together. It proceeds from something usable or exploitable. The principle normally proceeds from the effect or phenomenon (set of phenomena) it exploits. Energy efficiency can be considered as an effect. A technology that exploited nothing could achieve nothing: the phenomenon exploited need not be physical. It can use an effect from nature, a logical combination, a behavior or an organization. Buildings usually proceed from a number of effects to be combined and exploited, several principles of which may be adverse to each other. When energy efficiency in building is on the agenda it is often presented as if technology exploited a single, central effect.

A technology consists of a central assembly – this constitutes the backbone of the device, or method, that executes its base concepts and exploits one or more base effects – plus other assemblies suspended from this to make it workable and regulate its function. These components or assemblies function together in a working architecture or whole system (Arthur, 2007). Technologies are almost always adaptable in architecture, constantly changing in configuration and purpose as different needs require.

It is important to understand the principle and how this translates into components that share a working architecture. It is fundamental to recognize and make a clear distinction between principles and phenomena. A phenomenon is just a natural effect, while a principle is the idea of using this effect for some purpose. In our context, energy efficiency can be considered as a purpose, as a principle and as an effect. This confusion may actually become a barrier to learning. Changes in purpose, components or architecture normally imply a modified technology, but not a novel technology. A change in the base principle by which the purpose is achieved is a rare phenomenon and is a good candidate for a novel technology. A novel technology is then the one that achieves the same purpose by using a new or different base principle than used before.

Sub-problems arise as a certain component becomes necessary in order to achieve a certain principle or if a certain component is a pre-condition for a larger solution to follow. Each candidate principle brings up its own particular difficulties and these pose sub-problems. Principles are often borrowed, appropriated from other purposes or devices that use them and can be arrived at from two ends: from a need or from a phenomenon, with the difference that a principle can be sought from a need, but is suggested by the phenomenon.

In projects, there comes a moment of connection where the original problem is connected with a principle – an effect in use – that can handle it. The solution tends to be appropriate, elegant and as simple as possible. Seldom do projects in building arrive at that point, as decisions are made very early during the process. This process is recursive: it repeats until each problem (and sub-problem and sub-sub-problem) resolves itself into one that can be dealt with using existing components. It further becomes a building block, a multi-dimensional knowledge aimed at being used in future projects. What exactly are these building blocks used in projects? At a first sight, they are existing technologies in the form of components, assemblies or methods. Conceptually, in the project agent's mind, they are thought of as functionalities, generic actions or operations that lie at hand in order to shape temporary hypotheses. But knowing functionalities is not enough. What is of most importance is knowing what is likely *not* to work, what methods to use, whom to talk to, what theories to look at and 'above all of how to manipulate phenomena that may be freshly discovered and poorly understood' (Arthur, 2007; Polanyi, 1967).

Conclusions

In energy efficient building projects we are seeking a base principle, the idea of some effect or a combination of effects, in an action that will fulfill the requirements of the problem. Outcomes depend on the interactions between agents within the specific context of a project. 'Next time I'll make it another way, I have much clearer ideas now, and I know exactly what a better and simpler solution is.' But next time will be exactly the same. The subjective belief model interacting with other agents' belief models will adapt, change, co-evolve and make choices in that context, which once more will shape new belief models worth using and discarding others. The co-evolutionary aspect of learning in projects is important. Each agent's subjective beliefs evolve and condition the evolution of the beliefs of others.

On closer examination the projects may highlight ambiguous attitudes in the formulation of the purpose or of the basic technological principle. At a first glance, many of the projects may show particular focus on the effect to fulfill. Some of them use advanced and expensive components, while others try to simplify the technical solutions and the choice of systems as much as possible. All the variables at play have interactive effects for the evolution of projects and, as far as our experience goes, the fact that they are not exactly defined induces unusual conversational activities in the team – a pre-condition for learning (Wenger, 1998).

The project team is assumed to be the context within which learning about building and technology is generated. Different terms are introduced: problem, solution, technology, knowledge, hypotheses, models, purpose, principle and effect. These are terms for analysis; they are abstract terms, not used in project contexts, but as tools to identify the continuous passage from induction to deduction, drawing on intuition and behavior as much as on theory and rational analysis.

Problems of complication and ill-definition invariably occur in projects. Actors struggle within new formulations of goals and their own experience, their understandings – the associations and meanings they have derived from their history of previous actions and experiences. By copying from standard problems of building, we choose to ignore the issue. For the larger issue of sustainable development and building, we cannot – we have to take cognition seriously.

Acknowledgment

The author acknowledges the financial assisted afforded by the Swedish Research Council (Formas) and BIC to support the research reported in this chapter.

18 Participative design tools in inner-city redevelopment

Gert-Joost Peek and Jac L.A. Geurts

Introduction

Inner-city redevelopment processes are complex. They deal with both a physical as well as a functional transformation of a conjoint part of the city. It is about the construction of new real estate, infrastructure and public space while the area as a whole remains very much in use. Development and construction of these projects takes a long time and cannot be done by one entity on its own. In short, the complexity of inner-city redevelopment processes arises from the spatial as well as from the organizational context. Ambitions to turn these areas into multifunctional hotspots within the city are high, while the transformation itself means dealing with numerous stakeholders and restrictive environmental legislation and ensuring the continuation of transfer functions. In practice, we observe two starting points for these redevelopment initiatives. Initiators either start by commissioning the design of an alternative for the location or they start with an extended stakeholder analysis. Actors in the concept phase tend to cling for too long to one or the other approach. This creates problems. On the one hand, the design approach leads to a process of reacting to sketches, never arriving at a complete overview of stakeholders' demands and wishes. The stakeholder analysis approach, on the other hand, may well lead to a non-realizable pile of ambitions. Both approaches result in an unsteady course of action and lower the chances of enriching the design solution.

The authors propose a third way (Peek, 2006) that combines both approaches of design and stakeholder analysis with a participative design tool in order to intervene in the concept phase of the redevelopment process and contribute to a smoother process and the quality of the outcome. This chapter provides an overview of the literature on which both approaches are based and the background to the dynamic approach combining both. The design of the participative design tool and testing it are then described, followed by the results of those tests.

Literature review

Both approaches – design and stakeholder analysis – of the concept phase of inner-city redevelopment processes stem from two major schools of policy analysis. Traditionally, these processes are viewed as a succession of phases. In the early phases, emphasis is on variation, after which final choices are made. Mintzberg (1976) and Kingdon (1995) state that within variation the most (implicit) selection takes place. One can only select what is already designed or at least formulated. This has led to viewing these processes as 'garbage cans' (Cohen *et al.*, 1972). Variation and selection take place at the same time and are the results of a chaotic and accidental process. The normative structured view of a process in phases is replaced by such descriptive representations as the 'streams model' (Kingdon, 1995) and the 'rounds model' (Teisman, 1992; Teisman, 2000).

In the research presented in this chapter, we are looking for a combination of both schools of thought accepting the chaotic and accidental character of practice, but at the same time realizing that we need to structure the process in order to get things done. In the concept phase, when little is structured yet, the view of Kingdon's model of streams is adopted. The actual problem becomes clear when we confront this view with the dilemma of Collingridge (1980), which he observed in the introduction of new technology and which we find applicable to the concept phase of redevelopment processes. As the literature basis for our solution we turn to Corner *et al.* (2001), who have developed a conceptualization of decision problem structuring that synthesizes a number of models and approaches cited in the decision-making literature in general and the multi-criteria literature in particular.

Kingdon's model of streams

Kingdon (1995) envisions the possibilities for policy change (agenda-setting) when three, largely independent streams are coupled (see Figure 18.1). Combined actions between these streams include *problems* (denoting issues that are recognized as significant social problems), *policy* or, in our case, *alternatives* (referring to good advice), *political environment* (characterized by elections, changes in government and changes in public opinion) and the importance of chance, captured by the concept of *policy windows*, which makes policy change possible. Policy windows occur when there is an opening for new views and provide the opportunity to have alternative issues and solutions considered seriously. In short, critical factors in this model of agenda-setting are timing, chance and external influence. Problems and solutions may disappear or float to the top of the streams in a somewhat random manner, which means that important decisions can be taken in various places and with varying interest in the respective alternatives. The role of external influences also indicates that alternatives which are circulated within policy networks may have a significant impact when it has the chance to address an emerging issue at the right time and place. Bruil (2004: 272–6) found the concept of streams applicable to redevelopment processes.

Collingridge's dilemma

The dilemma of Collingridge (1980) hampers the occurrence of policy windows and makes using them difficult. Collingridge observed this problem in the early phases of technology development and we thought it applicable to the start of complex redevelopment processes. He formulates the dilemma as follows:

> Attempting to control a technology is difficult, and not rarely impossible, because during its early stages, when it can be controlled, not enough can be known about its harmful social consequences to warrant controlling its development; but by the time these consequences are apparent, control has become costly and slow.
>
> (Collingridge, 1980: 19)

While the possible alternatives are numerous, potential stakeholders – for the same reason – find it hard to imagine their effects. This results in a low level of participation by potential stakeholders at the start of redevelopment processes. This situation is especially true of those who could potentially benefit from participation; while at this time, the development of

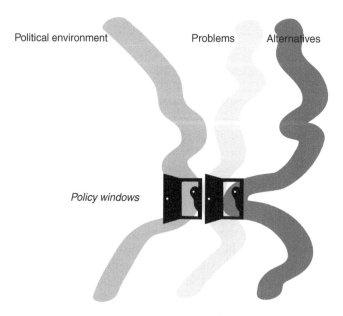

Figure 18.1 Model of streams (adapted from Kingdon, 1995).

alternatives – although still abstract – is very much open and controllable (Van Eijk, 2003: 199).

In Figure 18.2, Kingdon's model of streams and Collingridge's dilemma are in direct confrontation. Since stakeholders find it hard to imagine alternatives, they see little relationship between possible alternative solutions to the problem and their values. Policy windows stay shut, while the alternatives are very open to change and, consequently, there are many opportunities for aligning the three streams.

Kingdon tells us that it is essential for the start-up of redevelopment processes to study not only the spatial problem itself, but also to take account of the stakes of potential stakeholders – the political environment – as well as potential alternatives. Collingridge makes clear that, although stakes and possible alternatives are related, it is hard to establish this link in practice. There are so many directions in which alternatives could develop such that the actors involved cannot picture the possible effects of the redevelopment and find it hard to formulate their stakes.

Corner's dynamic approach

The predicament stemming from the confrontation of Kingdon's model of streams with Collingridge's dilemma are not new to the world of decision-making, especially when dealing with so called ill-structured or multi-criteria problems (Rosenhead and Mingers, 2001: 4–5, with reference to Ackoff, 1979; Rittel and Webber, 1973; Schön, 1987). Corner *et al.* (2001: 131, with reference to Keeney and Raiffa, 1976; Saaty, 1980) point out that most decision models are built from criteria and alternatives:

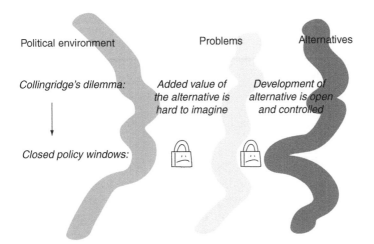

Figure 18.2 Confrontation of Kingdon's model of streams with Collingridge's dilemma.

Criteria reflect the values of a decision-maker and are the means by which alternatives can be discriminated. Alternatives are courses of action which can be pursued and which will have outcomes measured in terms of the criteria. Henig and Buchanan (1996) and Buchanan *et al.* (1998) present a conceptualization of the multi-criteria decision problem structure in terms of criteria and alternatives, with attributes as the bridge between them. More precisely, they state that attributes are the objectively measurable features of the alternatives. Therefore, the decision problem is structured so as to separate the subjective components (criteria, values and preferences) from the objective components (alternatives and attributes), with a view to improving the decision process.

When considering redevelopment processes as multi-criteria decision problems, two approaches to their structuring can be taken. First, criteria are identified in the decision-making process; next, alternatives are creatively determined in the light of such criteria and the choice is then made.

> Known as value-focused thinking (VFT), this has been advocated as a prescriptive, proactive approach. Here, the explicit consideration of values is offered as a starting point to the structuring process, and leads to the creation of opportunities, rather than the need to solve problems.
>
> (Corner *et al.*, 2001: 132)

Second, following March's (1998) view that values and criteria are formed out of experience with alternatives and suggesting decision simulations as a way to discover hidden values and, subsequently, promising alternatives,

> the process of first specifying alternatives in the problem structuring process, and then applying value and preference information to them in order to make a choice, is commonly referred to as alternative-focused thinking (AFT). It is clear from the descriptive decision-making literature that AFT is easily the more common procedure (see, for example, Nutt (1993)).
>
> (Corner *et al.*, 2001: 132)

Figure 18.3 shows both approaches. The distinction between them is not just academic.

> It is difficult to determine which does or should come first – criteria or alternatives – since both are vitally important to the decision problem structuring process. However, we would argue that the starting point is not the issue. What is important is that the decision-maker learns about one (criteria or alternatives) from working with the other. In a good structuring process, criteria and alternatives both do and probably should generate each other interactively.
>
> (Corner *et al.*, 2001: 132)

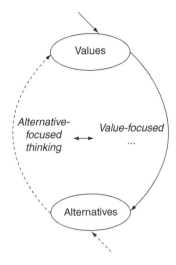

Figure 18.3 The alternative-focused and value-focused approach to multi-criteria decision problems (adapted from Corner *et al.* (2001).

The authors advocate a dynamic approach recognizing the different starting points inherent in VFT and AFT. More importantly, the approach reflects the interactive nature of criteria and alternatives, and suggests movement from one to the other. 'This interactive and dynamic approach for problem structuring implies that thinking about alternatives helps generate criteria, and vice versa. That is, neither of these two structuring elements can be thought of alone, independent of the other' (Corner *et al.*, 2001: 132–3). Corner *et al.* (2001: 135–8, with reference to Nutt 1993, 2001) show that the best results are generated by combining approaches based on alternative-focused thinking as well as on value-focused thinking.

Research project

The shortage of knowledge and experience in dealing with the design commission at the start of complex inner-city redevelopment processes has led to the development of a participative design tool aimed at creating coherence through the collaboration of long-term committed stakeholders in the concept phase. The point of departure is a new way of dealing with the objectives of the actors involved. These objectives should be 'knotted' rather than piled up. 'Knotting' encompasses a process for the initial actor and other stakeholders with a long-term commitment to the location to develop a shared view of the transformation based on a joint orientation toward possible synergies offered by the location. The design tool serves at least the following objectives/criteria:

- creation of mutual understanding by developing a shared language;
- recognition of mutual coherence and added value by analyzing objective–means relationships in day-to-day use in the location;
- search for 'win-win' situations by redefining individual ambitions into shared objectives;
- trade and adjustment of objectives by negotiation and compromise; and
- formulation of collaborative paths for exploiting synergistic opportunities.

Project description and objectives

The research project aims at adopting the dynamic approach for problem structuring to the concept phases of redevelopment processes. This method should facilitate a shared commissioning by actors who (jointly) have at their disposal a large share of the (financial) means required for the redevelopment. A focus on synergy is essential here, as it links the possibilities in the location and the objectives of the actor with a long-term commitment to the location. By this, a focus on synergy is a focus on quality in all its diverse aspects. Such is the view underlying the study. In this chapter, we focus on the method developed and not so much on the concept of synergy.

The design tool was developed and tested by an experimental implementation in the concept phase of the redevelopment process of Smakkelaarsveld, part of the station area of the city of Utrecht, being the fourth largest city in the Netherlands by inhabitants. The station is a major junction in the Dutch railway network. The master plan of the station area foresees Smakkelaarsveld as a gateway to the old city center and aims for a multifunctional brief including public functions, housing and a parking garage (Projectorganisatie Stationsgebied, 2003: 20). This ambition requires the participation of a number of actors. Table 18.1 lists those who participated in testing the design tool.

Design methodology

In order to create the design tool, the development of inner-city locations is viewed as an organizational process as well as a design process. In terms of design ambitions these processes are joined. Ambitions stem from the wishes of the actors involved, based on their objectives and means, and are bordered by the possibilities provided by engineering techniques. In a pluralist paradigm, actors need to find their common interests by participating in the design process.

Participation begs a common language consisting of notions dealing with elements of the location as a designer would use, with objectives of the stakeholders and with the relationships between these elements and

Table 18.1 Actors test the participative design tool

Long-term stakeholder	Interest
Project Organization Station area (POS) (municipality)	Progress of redevelopment
Bouwfonds MAB (developer)	Development opportunities
Kantoren Fonds Nederland (KFN) (investor)	Owner office building 'Smakkelaarsburcht'
FGH Bank (user-investor)	Owner-user office building 'Leidseveer'
Rabo Vastgoed** (developer)	Owner office space in shopping mall 'Hoog Catherijne'
Chamber of Commerce Utrecht (CCU)	Promotion commercial interest inner city and regional economy
NS Vastgoed* ** (developer)	Adjoining development

Notes
*no mind mapping.
**not attended workshop.

objectives. Systems theory offers the possibility of developing such a language by observing the inner-city location as a system. In the light of interwoven stakeholder interests and design enrichment, a central role is given to the synergic relationships between the elements and objectives. In systems theory, these ill-structured problems are dealt with by participative problem-structuring methods.

Participative problem-structuring methods are directed at mutual learning among actors involved in relation to the problem at hand. The essence of the method is a structured confrontation between the actors' images of the problem and possible alternative solutions. De Geus (1988: 71) views decision-making as 'a learning process, because people change their own mental models and build up a joint model as they talk'. 'Learning takes place when people discover for themselves contradictions between observed behavior and their perceptions of how the world should operate. So, managers must experiment with models, try their own *what-ifs*, and use simulations to trigger wide-ranging discussion' (Morecroft, 1992: 9). Mental models are 'networks of facts and concepts that mimic reality and from which executives [and indeed all people] derive their opinions of strategic issues, options, courses of action and likely outcomes' (Morecroft, 1988: 313). Mental models are constructed and constantly tested by people's daily dealings with reality and must be responsive to change. In time, mental models become implicit; they are within one's mind and are not directly available for discussion and analysis (Geurts and Vennix, 1989: 57).

The design tool is based on a dynamic approach to decision-making problems, i.e. an approach based on interests or values and an approach founded on solutions or alternatives which are used in turn. The value-oriented approach is represented by participative modeling utilizing an exercise in mind mapping through which the mental models of the actors are made explicit. The alternative-oriented approach by participative design is based on a design table which actors, working in small groups, use to design a possible alternative. Next, these alternatives are placed in opposition in order to get a common understanding about what the actors agree upon and the major differences between them. The result of this first sequence of viewing the redevelopment issue from the perspective of values and alternatives is not so much a design as the basis for a joint design commission that is able to foster coherence in the design and long-term collaboration by committed stakeholders.

Test results and industrial impact

The test of the design tool consisted of an individual interview and a joint workshop. During the interviews, mind mapping was introduced and the various actors were asked to indicate the most important factor for success and failure of the redevelopment.

Individual mind maps are constructed as follows. The actor begins by setting out all values that are influenced by the redevelopment, such as return-on-investment, architectural quality and safety. Next, attributes are added – features of the future location – and they are related to the values. In the last step, the relationship between attributes and values are scored from –9 (the attribute has a very negative effect on the value) to 9 (the attribute has a very positive effect on the value).

The actors were given a week to create their personal mind map and submit it online. The individual mind maps are consolidated into a collective model showing how each value is scored by all actors and how they regard the attributes influencing these values. At a plenary workshop, a few weeks later, the joint model based on the individual maps was presented and discussed. The actors then proceeded to work with a design table.

Results

Table 18.2 shows that the Project Organization Station assumes most values (6), followed by FGH Bank and the developer Bouwfonds MAB (5). The value 'return' and 'spatial and functional quality' are shared by most actors (4). The same values 'return' (12) and 'spatial and functional quality' (9) and the value 'operations' (8), referring to the phase of use, are influenced by most attributes. Under the heading of attributes, 'architecture', 'development urban quality' and 'routing pedestrians' influence most values (5). Through this and other analyses, actors gain an idea of the level of commonalty of their vision for the location. The presentation leads to interaction between the actors: they talk about the view they have from each other's position in the process and are able to adjust their perceptions.

Table 18.2 Connectivity of actors' values and attributes in the collective model of Smakkelaarsveld

→ Actor		← Value		→ Attribute	←
6	POS	4	Return	12	Architecture	5
5	FGH Bank	4	Spatial and functional quality	9	Development urban quality	5
5	Bouwfonds MAB	3	Continuity	5	Routing pedestrians	5
4	KFN	2	Operations	8	Parking space	4
3	CC Utrecht	2	Image and appearance	5	Development risks and opportunities	3
3	Rabo Vastgoed	2	Liveliness	5	Dwellings	3
		2	Accessibility	3	Functional mix	3
		2	Liveableness	2	Use of space	3
		1	Relations with surroundings	3	Accessibility (by car)	3
		1	Added value to city as a whole	2	Public space	3
		1	Multiple use of space	2	Buildings	2
		1	Parking	1	Urban plan	2
		1	Development	0	Commerce	2
					Amenities	2
					Green space	2
					Shops	2
					Added value to company	2
					Public transport	2
					Management of urban quality	1
					Image of the area	1
					Double use of parking space	1
					Sustainability	1

Notes
('→ actor' shows the number of values the actor has mentioned; '← value' shows the number of actors who have mentioned the value; 'value →' shows the number of attributes that influence the value; 'attributes ←' shows the number of values that are under the influence of the attribute).

After the participative modeling exercise, the actors started designing using the design table (based on the strategy table used by Geurts and Weggeman (1992)) in order to come up with alternative solutions. In the design table, the area's most important attributes are set out, based on the factors for success and failure as mentioned by the actors in individual interviews, followed by concrete solutions presented as options. The actors work in two small groups choosing one option for each attribute. In this way, an alternative for the location is developed that consists of a succession of the options chosen. Next, alternatives are tested for consistency and the extent to which they take an integral perspective. Subsequently, actors give their alternatives a catchy label. Figure 18.4 shows both

Purpose	Slow traffic connection	High-end public transport (HPT) line	Type of additional building(s)	Main function	Additional functions	Additional m² + functions	Architectonic expression	Definition of subarea 'Smakkelaarsveld'	Relation with other sub-areas of Station area	Time path: start of re-development	Leading parties	Type of partnership
	What?							How?			Who?	Extra design
Primary function 'Smakkelaarsveld'												Public–private partnership
Entrance to the city centre	No connecting route	No 'HPT' line	No additional buildings	No main function	No other functions	<1,000 m² offices and retail	Renovation existing buildings	Including existing facades	Related to station	Together with other sub-areas	Municipality	
Entrance to the offices	East–West rout: *biking*	No exclusive 'HPT' line	Solitary building at the rail-side	Library	Retail, restaurants, residences, offices	28,000 m² residential, offices, retail and *culture*	Historicizing, *upgrading*	Including offices to Knipstraat	Autonomous	Following 'Stationshal' and 'Stationstraat'	Real estate developer	
Entrance to 'Hoog Catharijne'	East–West route: *walking*	'HPT' line integrated in buildings	Solitary building at the city-side	Multiplex	Retail and restaurants	20,000 m² offices, retail and *leisure*	International style	Including offices to 'Daalsetunnel'	Related to offices	Following 'Vredenburg'	Investors	
Entrance to the station	North–South route	'HPT' line along buildings	Buildings added to existing walls	Large-scale retail	Residences and retail	20,000 m² residential, offices, *retail and public*	Avant-garde	Including connecting routes	Related to 'Hoog Catharijne'	Following 'Hoog Catharijne'	Potential users	
Solitary business function			Underground buildings		Offices and restaurants	20,000 m² *offices and retail*	Classical			Independent *subject to 'HPT' line*		
Other: *gate to city centre*	Other: ...	Other: *'HPT' line underground*	Other: *low density maintain open space*	Other: *custom-made*	Other: *custom-made*	Other: *15,000 m² offices, facilities residences*	Other: *contrast*	Other: ...	Other: *related to whole environment*	Other: *completion 2012*	Other: ...	
			Other: *buildings at rail side and city side*									

Figure 18.4 Design table example Smakkelaarsveld of group 1 (dark areas) and 2 (light areas).

alternatives in order to give the reader an impression of the structure of the table. Both alternatives are pictured as series of dark (group 1) and light (group 2) boxes.

Working with the design table forces actors to communicate their wishes and ideas and to negotiate over their objectives and willingness to contribute assets and means. The table offers them a window on the location's opportunities. As a final step, different alternatives are compared. This provides insight into the similarities and differences of alternatives and, by doing so, suggests directions where further steps should be taken.

A SWOT analysis revealed that the actors felt the design tool provided insight into the similarities and differences in their interests and generated a shared vision for the location efficiently and effectively. Moreover, it led to a discussion on the important design features and on negotiating stakes and assets. The method enabled the actors, despite the early stage in the process, to start searching for added-value opportunities collaboratively. The outcome of the design workshop could serve subsequently as a basis for the design brief when commissioning an urban planner or architect.

Implementation and exploitation

The design tool should contribute to a smoother and faster redevelopment process by coupling the objectives of actors involved through means of the location synergy concept and in doing so this will lead to ambitious, yet realizable, plans for creating added value in the operational phase for the location. Critical to achieving this goal is collaboration amongst actors who combine a long-term interest in the location with a sufficient means for realization of the plans. These long-term committed actors should not be sought solely from among those already involved in the location. In many cases, existing landowners and users will not have long-term stakes as they will leave the location and will not return after redevelopment. We should therefore search for actors with long-term interests matching those of the initiators and/or the location's potential. These actors should become involved in the redevelopment in its earliest stages.

The role of real estate developers is an interesting one. Although developers are not committed to the long term, their prime objective and conceptual and financial expertise equips them for the role of catalyst within these processes and, furthermore, positions them as particularly committed to progress. Actors lacking a long-term commitment to the location, who already play a central role in the policy arena crowding these processes, should be motivated to extend their time horizons. Next to this, shaping the plan's substance creates opportunities for stimulating long-term commitment. All these actions contribute to the formation of a group of actors who are able to benefit from a common direction focused on mutual added value.

Conclusions

Once long-term stakeholders have found each other, the next step is to get them collaborating in order to develop a coherent plan together that will generate added value in the operational phase. The research shows that an iteration of value-focused thinking and alternative-focused thinking is a solid basis for such collaboration and strengthens the prospects for successful redevelopment. In practice, actors in the concept phase tend to cling to one or the other approach for too long.

The participative design tool offers a mirror for viewing their own images of the future location and comparing these with the images of others. Moreover, the method provides a window on the location's potential and helps actors communicate their images and views on the location's potential. Finally, the method functions as a marketplace for negotiation to arrive at a consensus and commitment regarding the approach to the location. Still, the method does not guarantee a smooth and fast process or a synergy-rich result.

The design tool should not only result in the necessary conditions for the redevelopment, but the participative way of working should also contribute to the critical success factors of such processes. The method plays on the actors' expertise and professionalism. In the final analysis, these attributes prevail over those connected with formal positions or culture, thus establishing a dialog even in cases of mutual distrust. Situations of great urgency, when earlier attempts have failed, create the context for such breakthroughs. Once influential representatives of the actors are willing to commit themselves, the power of a participative way of working becomes apparent. In the context of the design tool, this power is combined with the strength of designing. This combination explains why redevelopments can take off despite unfavorable prospects based on scientific analysis of the background variables. A design-based intervention by a group of inspired and courageous people has the potential to change it all.

This research provides us with starting points for developing further the field of redevelopment. The design tool contributes to more professional commissioning by both public and private actors. The method challenges actors to look beyond their personal interests and means by appealing to their expertise and professionalism: are we not all in the trade of realizing outstanding projects?

19 Boundary objects in design

Kari Hovin Kjølle and Cecilia Gustafsson

Introduction

Over the past 15 years, the attention afforded to space and the physical environment in terms of office solutions has increased. New ideas about management, leadership, knowledge at work and the place of new technology, have led to a shift in how we think about the workplace. The context of work is changing; places and times of work and the way people interact are changing too. Similarly, the demand for highly flexible office space is increasing. Collocating staff, focusing on spaces for interaction and getting an invigorated and stimulating workplace culture (Vos *et al.*, 1997), is fostering teamwork, which adds value to the business and raises the brand of the companies concerned.

Changes have also led to a shift in the role of the architect in the building design process. The former master builder with total responsibility has been reduced to an actor among others in the briefing and design phases of a complex project, although architects still have a greater influence on the crucial conceptual design decisions during the design process. Even so, such concepts as teamwork and collaboration, as collective terms for every interaction with others, are not sufficiently precise to enable designers to interpret the clients' vision of a successful spatial solution. Hence, it is important to understand how architects and designers manage the supply side and the translation process from 'business language' to 'architectural language'. There is a chain of translations, where choices and decisions have to be made. The question is: can the translation process be facilitated by instruments that help the process go more easily and give the participants a shared and better understanding?

Based on a current case study,[1] we explore some instruments used in the translation process between the clients on the demand side and architects and interior designers on the supply side. In order to analyze how the collective action is managed across these social worlds, we have chosen a theoretical framework from the field of science, technology and society. We focus on the concept called 'boundary objects', which is useful for achieving sufficient agreement between stakeholders in order to progress decisions and get work done. The concept can be used to gain new methodological insight into the process of briefing and design (Kjølle *et al.*, 2005).

Literature review

A building is a social and material construction, where the order of space is the purpose. Buildings often become an object of cultural discourse, which complicates the relationship between usability and social meaning, and which implies that they are set apart from other objects and artifacts (Hillier and Hanson, 1984). Insofar as the building object creates and orders the empty volumes of space into a pattern, the building is also a 'black box'[2] for its users, representing the architects' and other designers' associations and viewpoints.

Boundary objects as a means of communication and translation

The development of facts and artifacts depends on communication between the participants, as well as the task of reconciling. This is the feature of the concept called 'boundary object' developed some years ago by researchers such as Susan Leigh Star, James R. Griesemer, Adele Clarke and Joan Fujimura. All of them are associated with symbolic interactionism, wherein the concept of social worlds has an important role. Social worlds are units of discourse limited not by geography but by communication, activities and processes.

The focal point is the collective work across different communities of practice, with divergent agendas and viewpoints. In other words, the focus of this concept is the multiple transactions needed and the equality between actors with divergent associations and agendas. The development of the concept of boundary objects was undertaken by Star and Griesemer (1989) through a study of different groups of actors crafting an easy and clear concept of problem solving, which made the groups intersect and work successfully together. They describe boundary objects as 'common enough to more than one world to make them recognizable' and define the concept as 'a means of translation' (Star and Griesemer, 1989: 393).

As interfaces between multiple social worlds, boundary objects are the means by which interaction and communication is affected at the places 'where people meet', described as a media of communication between the communities represented. They can be abstract or concrete objects that arise over time from durable cooperation and understood or misunderstood in equality between the participants. They are working arrangements, encouraging translation for the purpose of winning allies, which include allowing others to resist translation and to construct other facts. Creating and managing boundary objects are key processes in developing and maintaining coherence across different communities of practice and social worlds. Star and Griesemer (1989) identify certain criteria of the concept: first,

> a boundary object is an analytical concept of those scientific objects which both inhabit several intersecting social worlds and satisfy the informational requirements of each of them. Second, they are both plastic enough to adapt to local needs and the constraints of the several parties employing them, yet robust enough to maintain a common identity across sites. Last, boundary objects are weakly structured in common use and become strongly structured in individual site use.

According to Fujimura (1992), boundary objects are defined as entities at least apparently common to several actors' discourses, enabling them to discuss an issue and perceive a shared interest. She claims that boundary objects are an analytical concept 'that both inhabit several agreements at various times to get work done and produce relatively and temporarily stable facts' (Fujimura, 1992: 168). Furthermore, Fujimura points out that boundary objects 'facilitate the flow of resources' such as concepts, skills, artifacts, materials, techniques and instruments among several different lines of work. Their inconstancy is the strength of the concept and they assist with the need for cooperation without accordance. Clarke and Fujimura (1992) define boundary objects to include things, tools, artifacts and techniques, in addition to ideas, stories and memories – 'objects that all are treated as consequential by community members'.

In the Norwegian R&D project, KUNNE,[3] the concept of boundary objects has been used in order to describe 'objects that become shared foci for the attention and explorative activities of people with initially different interests, expertise and language' (Carlsen *et al.*, 2004: 229). The importance of the loosely structured nature of the objects has been highlighted, allowing participation in development and construction of the boundary object.

Other aspects

How communication processes and organizational design react and work in virtual and global organizations or companies are aspects which have a large influence on the communication between the demand side and the supply side in the translation process in briefing and design. Communications become more task-oriented with clearer role expectations when problem solving between actors is computer mediated, while face-to-face communications are more cohesive and personal.

A cluster analysis of communication patterns by the authors has shown that as computer-mediated group interactions progress, group decisions and general problem-solving processes became more closely related. In contrast, the study showed that face-to-face group interactions often followed sequences of interactions, because the participants reflected more on perspectives and courses of action in reaching their decisions. One finding in the study was that actors, who solved problems through virtual meetings, were more satisfied and believed there was a greater quality in the problem-solving process (Jonassen and Kwon, 2001).

Research project

The study presented in this chapter is exploratory, aiming to define boundary objects used in the design process as a means for translating needs into design. The boundary objects discussed are primarily used by architects and interior designers on the supply side.

Project description and objectives

The study was conducted at the headquarters of Office Design,[4] an international office design and architecture company. The company, which was founded in Europe in the 1970s, has offices in Europe, North America and Australia and had approximately 250 staff worldwide at the time of the study. Some 60 staff were based in the office under observation.

The company's profile is to demonstrate business values through efficiency (reducing costs), effectiveness (impacting productivity) and expression (impacting brand value). The idealism of the founders was an attempt to have a designer culture in parallel with a consultancy culture. Over the years, the company has developed a solid reputation for developing innovative office solutions, based on research and scientific inquiry. The office had been designed in the late 1990s as an example and showcase for the company's philosophy of open settings and flexible working. The services offered by the company range from urban planning to interior design, from a base of strategic briefing and research.

The company was chosen for the case study because of its integrated design approach, which focuses on the process, going from understanding the business objectives to understanding how people work, how space is used and what role it plays in the organization. This approach requires an ability to translate business needs into spatial solutions, as well as working in cross-disciplinary teams, which are the two main foci of this research project.

Research methodology

This study is aimed at exploring human experiences in detail from a qualitative perspective; thus, a constructivist approach was chosen, whose subject matter is that reality is socially constructed. The task is to understand the multiple social constructions of meaning and knowledge. Hence, a case study methodology was adopted (Robson, 2002; Yin, 2003), where such methods as interviews and observations have allowed us to acquire multiple perspectives. The researcher was a participant observer (Yin, 2003; Robson, 2002), working within the company for six months. During this time in three projects, interactions and collective work were mostly observed and analyzed, and several interviews conducted with actors from different disciplines and social worlds.

Data were collected through multiple methods and sources of evidence. In addition to interviews and observations, documents were studied and archival records collected. The interviews have been semi-structured, following an interview guide with open-ended questions. It was then possible to triangulate the data collected, draw comparisons and use the data in a complementary fashion to enhance interpretability (Robson, 2002).

The projects and interactions

Three very different projects were studied, all in progress during the study. The first project, called Global Pharmacy, had a high degree of user involvement. The client was demanding, in terms of being prepared and proactive. A pilot hub recently produced in the USA became a kind of guideline in addition to the research undertaken by strategists and researchers primarily from Office Design in the USA. The request was for the design of five similar hubs, in existing buildings, in two countries. They were intended to be new hubs for interaction, encouraging connections between scientific and creative people from diverse departments (see Table 19.1).

There was less user involvement in the second project, Fast Office, a rapidly expanding ICT company. The request was for an open-plan design solution, focusing on spaces for interaction, in two existing buildings, for their main office in the UK and their European head office in Ireland. Furthermore, the demand was 'fun, function and food'. The designers had experience from several projects delivered for this client in different countries and this made it easier for them to understand the client's needs and expectations. The last project observed, Science Park, was a project initiated by a developer and had no user involvement. The need was for the design of a science innovation center for commercial use. The request was vague and formulating the brief became a part of the delivery from the architects in the project.

Observation process

Four groups of stakeholders in the design process have been observed: strategists, researchers, interior designers and architects – the last two were the primary groups. The stakeholders took part in different stages of projects, with periods of overlapping communication and activities. They came from different educational backgrounds and thus had different preferred

Table 19.1 Observed interactions

Type of communication	Physical meetings client/ focus groups and designers	Virtual meetings client/focus groups and designers	Between designers and other actors	Between designers internal
Global Pharmacy Global Pharmacy: two hubs in the UK	Face-to-face communication in physical meetings	Virtual meeting	Face-to-face communication: construction architect and construction team on site	Interaction, collaboration between designers. Design review internal
Global Pharmacy: three hubs in the USA		Virtual meeting	Virtual meeting with the project team including the local architect	Interaction, collaboration between designers. Design review internal. A strategist ('bridge-builder') invited
Fast Office Fast Office Ireland	Some face-to-face communication in physical meetings	Virtual meeting	Virtual meeting with the project team including the local architect	Interactions between two designers, ad hoc and of short duration
Fast Office UK	Face-to-face communication in physical meetings	Virtual meeting with the project team, including the local architect	Face-to-face communication: furniture supplier – construction team on site	Interactions between two designers, ad hoc and of short duration
Science Park on the west coast of England	Face-to-face communication in physical meetings		Face-to-face comm unication in physical meetings with the project team, including the local architect	Interactions within the architect's team: discussion, interpretation and understanding of place (site and location), strategic brief, the concept of space plan and concept of the building

communication tools and patterns. Furthermore, individuals often change professional roles within the company (designers becoming strategists, for example) as well as within projects. This added an extra layer of complexity to the already multi-professional environment of the projects.

For exploratory purposes, the observation study of the interactions initially took an unobtrusive approach and became purely ethnographic (Robson, 2002). As the study progressed, interactions increased between the observer and the staff observed, which took it to the level of participant observation. The researcher did not attempt to influence the design process and results. Interaction was mainly in terms of how the design process was run with a focus on the role of the 'boundary objects' and the communication between the actors. According to Hastrup and Hervik (1994), fieldwork has to be experienced as performed. The ethnographic material is composed of more than words, which made the researcher aware of the paucity of the language used to characterize it.

Observed meetings were both formal and informal, and the interactions face-to-face as well as through the use of ICT (e.g. conference calls). In addition to internal meetings, there were meetings with clients, focus groups, project teams and furniture companies – all have been observed. Some of the meetings or interactions were recorded on film, the rest were captured in notes.

Interview process

Thirty semi-structured interviews (Yin, 2003; Robson, 2002) were conducted with members of the four stakeholder groups within Office Design. Five interviews were conducted with representatives of the clients in the projects under observation and one interview with an external project manager. Every interviewee had an interview guide in advance, together with a slideshow presentation (printed version) of the various instruments that the strategists and interior designers would use.

Handing out the interview guide in advance was meant to help the interviewees stay focused and to make the atmosphere more relaxed so that conversation would be open to exploration of different paths when covering the questions. Maintaining focus also helped to achieve a deeper discussion, and along unforeseen paths, resulting in richer data collection and growth in the study as a whole (Gubrium and Holstein, 1997; Kvale, 2001; Robson, 2002; Yin, 2003). Questions concentrated on instruments in use, communication, cooperation and knowledge sharing in general in projects, as well as between colleagues and across teams internally.

Different ways of addressing the focus and the topic were chosen in the interviews, depending on a number of factors, primarily the identity of the interviewee and the two actors' schedule for the day. The interviewees became narrators or storytellers. As the interviews went along, the stories were improvised in a meaningful talk by volunteering aspects of experiences, opinions, feelings and expectations. By working together and adopting an active view of the interview, a greater range of interpretive activities for both parties might emerge such that the interviews became social productions. Active interviewing is inherently collaborative, but also problematic (Holstein and Gubrium, 1995). Divergent social realities were conveyed, while understanding was shared and expressed during the conversation between the actors with a common interest in the topic. Meaningful reflections of, and personal attitudes about, the topic were brought out not just by words, but by accents, choices of words and body language. The researcher used herself as an instrument, with body language as well as words, to create an emotional interpretation to assist in comprehension (Kvale, 2001; Gubrium and Holstein, 1997; Yin, 2003).

Instruments as a means of translation and communication

Two factors contributed to its success as an internationally-leading consulting firm for designing office solutions. These are the development of methods, models and concepts, and the instruments we can define as boundary objects.

Instruments developed and standardized

For more than 30 years, by focusing on the importance of the brief and program to bring about a meaningful design, a series of methods, techniques, concepts, models, tools and artifacts have been developed. This collection of instruments has been further evolved, depending on the individual nature of the instruments, by those who have used them. Over time many of the instruments have become an integral part of the services and project management system in the office.

Rival firms, and others in the construction industry, have adapted and implemented some of the instruments developed in the office. An example instrument is Building Appraisal, which arose from a simple idea and quickly became a tool that was later implemented in the industry to compare, rate and rank buildings. The instrument is intended to validate the developer's brief and program through a series of critical evaluations of a building, matching the occupiers' business visions to its building stock and validating the architect's design against the users' brief, acting as a consumer advocate during the design process.

Four types of boundary object

We have found instruments that are a means for translation, as well as a means of communication. In analyzing the translation tasks, we can distinguish four types of boundary objects: repositories, standardized forms, ideal types and coincident boundaries (Star and Griesemer, 1989). In the study, four categories of staff are of interest in terms of their use of these boundary objects. The tables below give the initials of the categories of staff involved: researchers (R), strategists (S), interior designers (D) and architects (A). The strategists are actors with different skills and backgrounds.

Repositories

Repositories, in the sense of boundary objects, are ordered 'piles of objects'. They are archived or arranged in a standardized manner, which means they fit problems of heterogeneity across the communities. Project documents distributed amongst the stakeholders are of such a type, allowing individual actors to use them without having to negotiate directly over differences in purpose. Distributed and archived project documents, artifacts, drawings, texts and instruments are typical objects for actors. Books written or co-authored by some of the founders and staff in the office, on topics as the briefing process, architectural knowledge, the distributed workplace and new office solutions, are boundary objects of this type. Pictures on the walls of offices and drawings of building appraisals are repositories too.

   An important boundary object of this type is the story and the idea of the office, which were founded on the expectation that the two cultures of design and research can complement and support one another. The ethos of the office was built on responsibility, in terms of being a field containing understanding of sharing and risk, crossing social worlds and being forced out of the frame of reference. Since the office was founded, many instruments have been applied and have evolved in collective work amongst staff from different disciplines and social worlds as a natural aspect of understanding the purpose of a particular design.

Standardized forms

This type of boundary object is intended to be a method for common communication. The instruments can be transported over a long distance and retain the same information. We found two instruments of this type used by the researchers (R) and the strategists (S) early in the process of formulating the brief (Table 19.2). These are surveys about work patterns and use of space. The results are presented in meetings, communicated and discussed in an iterative briefing and change process with the leadership, the board, senior members of staff or focus groups. Slide presentations are developed, with illustrations, graphs and models that satisfy the informational requirements of the diverse stakeholders and inhabit their intersecting social worlds. A common understanding is maintained during the discussions.

Table 19.2 Standardized forms

Instruments	What	The issue	By whom
Work pattern survey	Workplace Performance Survey; Work Pattern Survey; Mobility Survey Communication	Collecting data work patterns	R/S
Utilization survey	Time Utilization Survey; Space Utilization Study/ Entity Analysis	Measuring how space is used, the time it is used for or the kind of staff in observed places	R/S

Ideal types

Building types, office types, diagrams, maps or other descriptions which, in fact, do not exactly describe the details of anything or one area, and which are abstracted from all domains and which may, to a large extent, be vague are defined as an ideal type in the sense of a boundary object. They can be regarded as adaptable, since they are able to guide all divergent actors equally, i.e. these instruments enable the actors to reach consensus. For this reason, ideal types can be defined as a means for communication and cooperating symbolically – see Table 19.3.

The analytical tools, diagrams and sketch drawings are presented in meetings as options, which facilitate discussions and negotiations. When agreements are made, these types of boundary objects are further developed for new meetings in an iterative manner. The instrument 'space planning' has more distinct steps, in which consensus about relations and logistics vertically must be reached before the next step (about models of relations horizontally) can be discussed.

Coincident boundaries

Common objects having the same boundaries, but with divergent internal contents, are another type and termed 'coincident boundaries' – see Table 19.4. They arise in the presence of different means of aggregating data, which implies the objects live in multiple social worlds and have different identities, contents and associations.

Visioning sessions and other types of workshops with cards, jigsaws or emotional tools are types of coincident boundary. In the course of regular meetings, discussions and negotiations between divergent stakeholders are facilitated by further developed slide presentations, which help them gain a common understanding of wants and needs, strategies and requirements. In addition to formulating the brief, these types of instruments help in the process of transforming the request into design. Later, in the design phase,

Table 19.3 Ideal types

Instruments	What	The issue	By whom
Targeted insight	An ethnographic study of employees' use of space. Micro study: individual use. Macro study: spatial process mapping. Campus: connectivity mapping	Study of the relationship between the environment and employees' behavior. Often used in labs and complex working areas, to get a targeted insight of space used	R/D
Space planning	Test of, and advise on, the suitability of (a range of) buildings. Translate the space model into their 'stacking' and 'blocking' tools	Space budget, alternative solutions showing the sizes of needs etc. Models of relations vertically then horizontally	R/D
Diagrams	Relationship diagrams, connectivity diagrams, graphs and 3D visualizations	Showing the relationship with, or connectivity to, other buildings, other elements etc	A
Analytical tools	Statistical tools, building types etc	Show sizes, dimensions etc	A
Sketch drawings architecture	Hand drawn or CAD drawn 2D and 3D sketches of site and building, annotated on paper, a whiteboard or flipchart	Showing the development of the concept and idea of the building	A
Sketch drawings interior	Often hand drawn 3D sketches on paper of interior elements, furniture, walls, colors, mood etc	Sketches showing the proposed (or some options of) interior details, furniture, rooms etc	D

tactile instruments such as physical models and boards contribute to equality among the stakeholders by showing the site, buildings and real materials. Hence, a common understanding of the proposed forms, space, sizes, colours and materials is enhanced. Another coincident boundary is their own office, which is a tactile instrument, serving as a showcase in action of an open-plan solution with new ways of working. On the walls are images and pictures, sketches and drawings of sites, buildings and interiors, which can be defined as several types of boundary object to facilitate a better and higher level of common understanding.

Table 19.4 Coincident boundaries

Instruments	What	The issue	By whom
Visioning session	Often series of workshops. Slide presentations with data collected and results from opinion surveys or interviews of boards and senior members of staff	Formulating the brief	D(S)
Workshop with cards	Pictures and images for drawing the needs and requirements	Particularly to formalize the strategic vision by the change process. Formulate the brief	R/S/D
Workshop with jigsaw	3D sketches of interior with elements that can be used in the office	Inspiring the client to define what they want in their new office. Formulating the brief	R/S/D
Emotional tools	Pictures and images of landscape and buildings. Examples of buildings' exteriors: form, expression, materials; examples of interiors: mood and atmosphere, contemplated spaces; metaphors. Visiting objects	Inspiring the client to define what they want in their new office. Contribute to the emotional 'cocktail'. Formulate the brief	A/D
Physical model	3D model of the building and the landscape	Show the form and size, the relation between the building and the landscape	A
Board	Board with pieces and examples of materials proposed, colors and/or surfaces	Show the 'real' colors, materials and surfaces proposed	D
The office	Open plan solution with partly nomads working on laptops, partly fixed workstations	Show new clients and visitors new ways of working	All staff
GBU – The Good, the Bad and the Ugly	An informal knowledge sharing session organized every fortnight for all staff	Sharing knowledge across teams and disciplines. Experiences and skills presented and discussed	All staff

Impact of boundary objects and time

For all three groups of projects, whether there was high, low or no user involvement, the process and time taken contributed to achieving high impact from the use of boundary objects. Sequential and iterative processes have, over time, a strong effect on understanding across the communities, clarifying needs and requirements, briefing, interpretation and implementation.

Boundary objects in the form of slide presentations developed further after each meeting and became strong instruments facilitating the discussions, influencing an increased, common understanding between participants enabling them to achieve agreements which again helped the translation process. The clients of the projects, Fast Office UK and the five hubs for Global Pharmacy, are very satisfied with the results and stress the impact that the slide presentations had on their understanding and the dialog they had with the interior designers.

Another finding is that many of the actors on both sides seemed to be more satisfied with face-to-face interactions than virtual meetings. The nature of the process, when engaged in regular meetings, opened up the opportunity to share reflections as well as enhance the relationship. Using more time, but being together physically to observe each other's body language, made the participants experience trust and a higher degree of loyalty to the group – more than was possible through virtual meetings. They felt they developed a better understanding and achieved a successful result.

Collective work and knowledge sharing

The other factors contributing to the impact of the instruments are actors, individuals and social worlds. As for instruments being boundary objects, there are differences in understanding over the effect of the various instruments, depending on levels of individual skill and discipline. In other words, the impact of a boundary object depends on the skill and understanding of those using it. 'It was an unwritten rule, the tool had a collective ownership', according to one the designers interviewed.

Over the years, the collective development of the instruments in the office has been one of the strengths. Through projects, as well as from one project to another, strategists and designers have worked together in teams, and senior members of staff have mentored junior members, all with the purpose of understanding the relationship between the demand and the supply sides, the performance of instruments, and making a better design. Now, the collective work across the disciplines depends on a group of communities.

A few people, in the cross-disciplinary strategist team including the researchers, are sharing and shaping knowledge and developing instruments together in an open way. There is some knowledge sharing and working between the strategists and the interior designers, but only a few interior designers knew about which instruments the strategists used. Between architects and strategists, there is less cross-working and little knowledge sharing. Between architects and interior designers, there is no cooperation and very little knowledge sharing. Meanwhile, staff share knowledge virtually through emails, which seems to depend more on individuals than on social worlds. Unfortunately, the amount and quality of emails is rather difficult to measure.

Conclusions

We have seen that boundary objects act as temporary bridges during the translation process, having a high impact on autonomy and communication between stakeholders from different worlds. The briefing and the design processes where design decisions are taken depend on the values and interests of the actors involved. The way individuals look at a problem in design discussions is related to beliefs and knowledge about the nature of good design practice, mixed with expertise, experience and responsibility. The tendency, as we have seen in the study, is for people to be more specialized. This means less sharing and less cooperation in work. People are very connected to their computers and sensitive to time pressures and timesheets. There is 'no place on the timesheet' for knowledge sharing. An important question raised by this study is how cooperation coexists while the diversity of the constituent parts remains significant in the translation process. A further question concerns actors: how can they act to contribute to the process of translation? Future work will investigate the answers.

Acknowledgments

The authors gratefully acknowledge the Norwegian Research Foundation and the Norwegian University of Science and Technology for their financial support. We thank the people at Office Design for making our empirical study possible.

Notes

1 This case study is a part of a doctoral research project which is related to the Norwegian multi-disciplinary R&D project 'The KUNNE workplace' (KWP). The project is a research initiative focused on new office solutions and new ways of working in knowledge-intensive organizations.
2 'Black box' is a term coined by cyberneticians for a piece of machinery or a set of commands that proves to be too complex.
3 The conception of knowledge in the KUNNE is grounded on a humanistic tradition. Knowledge results from the interaction between people, and between people and technology.
4 To maintain confidentiality, Office Design is used instead of the organization's real name.

20 Methodological and other uncertainties in life cycle costing

Erika Levander, Jutta Schade and Lars Stehn

Introduction

Life cycle costing (LCC) for buildings is an important tool for involving the client in the early stages of design where decisions affecting quality and cost over the long term are determined. Despite its acknowledged importance, life cycle costing has found limited application (Bakis *et al.*, 2003). As an example, an office building will consume about three times its initial capital cost over a 25-year period, but still more attention is paid to the initial or capital cost (Flanagan and Jewell, 2005). As demonstrated by Kotaji *et al.* (2003), it is particularly important to show the relation between design choices and the resulting lifetime cost (i.e. energy, maintenance and operations). Many clients have given a low priority to LCC because they are unaware or unconvinced of its benefits. From the client's perspective, there has to be a clear output motive for using LCC techniques (Cole and Sterner, 2000). Initial cost can be determined easily and reliably, but maintenance and operational cost data are not widely available. Where they are available, they can be less predictable since they must extend into the future. The limited availability of these data can prevent aspiring clients from taking a long-term perspective such that initial cost remains the main basis for decision-making. The absence of a framework for collecting and storing data is given as one cause for this lack of data (Bakis *et al.*, 2003).

Industrialization of construction is often put forward as a solution to concerns over quality and cost, and a means by which future costs can be better controlled. None of this thinking is new, although interest seems to have grown significantly over the past few years. Multi-dwelling timber-framed housing has been identified as one area for further industrialized process development, particularly in Sweden and other countries where there is a tradition of this form of construction. Against this encouraging background, uncertainty has been expressed by clients and building owners over long-term costs, technical performance and the management of prefabricated, timber-framed housing (Höök, 2005). A particular concern is that the industry's project orientation can mean the lack of a systematic and strategic approach to change (Saad *et al.*, 2002). Since project orientation and project culture are endemic, a method that incorporates uncertainties caused by, for example, industrialized housing must be able to work on the project level and, at the same time, stimulate long-term interest in the product. A weakness is seen in the reliability and uncertainties in the methodology of whole life costing.

This chapter discusses the limited application of LCC in construction from the perspective of clients and building owners and identifies the advantages and disadvantages of the main theoretical economic evaluation methods for LCC calculation as well as the sources of relevant data. The results of an interview study with building owners are presented, where the objectives were to identify uncertainties surrounding the greater adoption of timber-framed housing and to investigate if LCC could be used to address those particular uncertainties. In the context of this chapter, timber-framed housing is the industrialized production of multi-storey, multi-dwelling houses.

Background

Among contractors there is a specialization trend toward an increased use of prefabrication and industrialization in housing construction. As far back as 1995 in Sweden, the building regulations adopted a functional view that allowed timber to be used in multi-storey buildings. Of special interest for this research is, therefore, the development of timber-framed housing, in particular timber volume element (TVE) prefabrication. TVE is examined because it exemplifies several of the attributes that are important in the industrialized approach to timber framed construction.

TVEs are prefabricated as ready-to-use housing volumes complete with electrical installations, fittings and finishes. Prefabrication is a competitive option in the detached house market, but much less in the multi-storey market (around 10 percent of market share). The reason for this low acceptance of TVE prefabrication was examined by Höök (2005), through a survey of the attitudes of 35 building owner organizations, and produced the following results.

- *Historical prejudice.* TVE prefabrication was associated with barracks and mobile homes and, hence, with a historical prejudice about poor performance and low quality.
- *Lack of required technical information.* Technical solutions of TVEs and timber in itself were not thought to be able to fulfill all code-based functional demands such as adequate sound insulation.
- *Low long-term economic performance.* The TVE building system and manufacturers' capacity, as well as intentions over long-term quality and life cycle costs, were questioned.
- *Organizational or project management change.* TVE management is

related more to process than traditional project management. This necessitates new forms of cooperative relationships between the client and manufacturer.

Hence, the presumed beneficial effects of industrialization seem to be limited in the case of TVE prefabrication due to two combined effects. First, organizational and technological changes seem to outmode traditional construction management practices and place greater emphasis on coordination between the organizations. The client has to take a more active role in coordinating the industrialization process and probably lacks know-how and/or confidence in whether or not the TVE system will produce an optimal life cycle performance. Second, in the eyes of clients, there are too few organizations in the market to ensure reliability of the TVE system comparable with other construction technology.

Literature review

The primary literature covers life cycle costing, whole life costing (WLC) and whole life appraisal (WLA) and was reviewed in order to understand both definitions and usage in the construction industry. Integrated life cycle design was also investigated for its particular applicability. Additionally, literature on transaction costs was reviewed to appreciate the basis of the uncertainties referred to in the former literature, insofar as it concerns client interests.

Definition of LCC, WLC and WLA

Traditionally, the focus in construction is on minimizing the initial cost of the building. It has long been obvious that it is unwise to base the choice between alternatives solely on initial cost (Kishk *et al.*, 2003). This philosophy is today denominated differently by various authors. The terms used in the literature are 'cost in use', 'life cycle cost' (LCC), 'life cycle cost analysis' (LCCA), 'whole life costing' (WLC) and 'whole life appraisal' (WLA). Flanagan and Jewell (2005) report that terminology has changed over the years such that 'cost in use' has become 'life cycle costing' and also 'whole life costing'. They refer to the comparatively recent term, WLA, which embodies consideration of the cost, benefits and performance of the facility/asset over its lifetime, defining it as 'the total cost of a facility/asset over its operating life including initial acquisition costs and subsequent running costs'.

ISO 15686–5[1] (ISO, 2008) differentiates between the expressions WLC and LCC. Their contention is that WLC is equivalent to LCC plus external costs, thereby defining WLC as a broader term including, within it, life cycle costing and a broad range of analysis. Even in the ISO standard, there is the admission that sometimes these terms are used interchangeably, but the standard does try to interpret these terms more narrowly. The standard differentiates between LCC and 'life cycle costing' by stating that LCC should be used to describe a limited analysis of a few components, whereas 'life cycle costing' should be understood as the cost analyses. The Norwegian Standard 3454[2] (NS, 2000), for instance, defines LCC to include both initial costs and cost incurred throughout the whole functional lifetime and its ultimate demolition.

Discussions about wording can easily confuse. For the purpose of this chapter, LCC is used as the equivalent of WLC. LCC analysis is, in this context, to be understood as a broad analysis over the whole life cycle of a building. The term, LCC, was chosen as it is still the better known term in practice today.

Methods of LCC

For input to the LCC calculation, future costs are converted to their present equivalent using a suitable discount rate. A period of analysis is chosen and an appropriate economic evaluation method is applied. The literature reveals a wide variation in economic evaluation methods for LCC analysis. All have advantages and disadvantages. The methods have been formulated for different purposes and the user should be aware of their limitations. The relevant literature is organized in Table 20.1 and illustrates the six main economic evaluation methods for LCC, their advantages and disadvantages and purposes for which they can be used. The literature shows that the most suitable approach for LCC in the construction industry is based on the net present value (NPV) method. According to Kishk *et al.* (2003), the NPV method is also the most widely employed.

A review of different LCC models revealed that most of them use the same basic equation. What separates them is, however, the breakdown of cost elements. In terms of their suitability for construction, each model appears to have some particular advantages and some disadvantages (Kishk *et al.*, 2003). The model from the American Society for Testing Materials (equation 1), for example, distinguishes between energy and other running costs, which is useful in adopting different discount rates for different cost items.

$$NPV = C + R - S + A + M + E \ldots \tag{1}$$

Where C = investment costs; R = replacement costs; S = the resale value at the end of study period; A = annually recurring operating, maintenance and repair costs (except energy costs); M = non-annually recurring operating, maintenance and repair costs (except energy costs); E = energy costs.

Since LCC, by definition, deals with the future and the future is unknown, a risk analysis should be carried out after any calculation has been performed.

Table 20.1 Advantages and disadvantages of economic evaluation methods for LCC

Method	Approach	Advantage	Disadvantage	Applications
Simple payback	Calculates the time required to return the initial investment. The investment with the shortest payback time is the most profitable (Flanagan et al., 1989).	Quick and easy calculation; the result easy to interpret (Flanagan et al., 1989).	Does not take inflation, interest or cash flow into account (Öberg, 2005; Flanagan et al., 1989).	Rough estimation to see if the investment is profitable (Flanagan et al., 1989).
Discount payback (DPP)	Basically the same as the simple payback method, it just takes the time value into account (Flanagan et al., 1989).	Takes the time value of money into account (Flanagan et al., 1989).	Ignores all cash flow outside the payback period (Flanagan et al., 1989).	Should only be used as a screening device, not as a decision advice (Flanagan et al., 1989).
Net present value (NPV)	NPV is the result of the application of discount factors, based on a required rate of return to each year's projected cash flow, both in and out, so that the cash flows are discounted to present value. In general, if the NPV is positive it is worthwhile investing (Smullen and Hand, 2005). But as the focus in LCC is on cost rather than on income, the usual practice is to treat cost as positive and income as negative. Consequently, the best choice between two competing alternatives is the one with minimum NPV (Kishk et al., 2003).	Takes the time value of money into account. Generates the return equal to the market rate of interest. It uses all available data (Flanagan et al., 1989).	Not usable when the alternatives being compared have different life times. Not easy to interpret (Kishk et al., 2003).	Most LCC models utilize the NPV method (Kishk et al., 2003). Not usable if the alternatives have different life times (Flanagan et al., 1989).
Equivalent annual cost (ECA)	This method expresses the one-time NPV of an alternative as a uniform equivalent annual cost. For that it takes the factor present worth of annuity into account (Kishk et al., 2003).	Different alternatives with different life lengths can be compared (ISO, 2004).	Just gives an average number. It does not indicate the actual cost during each year of the LCC (ISO, 2004).	Comparing different alternatives with different life times (ISO, 2004).
Internal rate of return (IRR)	The IRR is a discounted cash flow criterion which determines an average rate of return by reference to the condition that the values be reduced to zero at the initial point of time (Moles and Terry, 1997). It is possible to calculate the test discount rate that will generate an NPV of zero. The alternative with the highest IRR is the best alternative (ISO, 2004).	Results are presented in percentages which provides an obvious interpretation (Flanagan et al., 1989).	Calculations need a trial and error procedure. IRR can only be calculated if the investments generate an income (Flanagan et al., 1989).	Can only be used if the investments generate an income, which is not always the case in the construction industry (Kishk et al., 2003).
Net saving (NS)	The NS is calculated as the difference between the present worth of the income generated by an investment and the amounted invested. The alternative with the highest net saving is the best (Kishk et al., 2003).	Easily understood investment appraisal technique (Kishk et al., 2003).	NS can only be used if the investment generates an income (Kishk et al., 2003).	Can be used to compare investment options (ISO, 2004), but only if the investment generates an income (Kishk et al., 2003).

Data for LCC

Data requirements for carrying out LCC analysis are, according to the literature, categorized in Figure 20.1. These different data influence LCC at different stages of the life cycle.

Occupancy and physical data could be seen as the key factors in the early design stage. LCC estimation in this stage depends on data such as floor area and the general requirements for the building. Flanagan *et al.* (1989) stressed the importance of occupancy data as equally key factors, especially for public buildings. Performance and quality data are rather influenced by policy decisions such as how well the buildings should be maintained and the degree of cleanliness demanded (Kishk *et al.*, 2003). Quality data are highly subjective and less readily accountable than cost data (Flanagan *et al.*, 1989; Flanagan and Jewell, 2005). In the more detailed design stage, life cycle cost estimation is based on performance and cost data for the building (Bakis *et al.*, 2003). Cost data are essential for LCC studies. Cost data that are not complemented by other types of data would be almost meaningless (Flanagan *et al.*, 1989). These data need

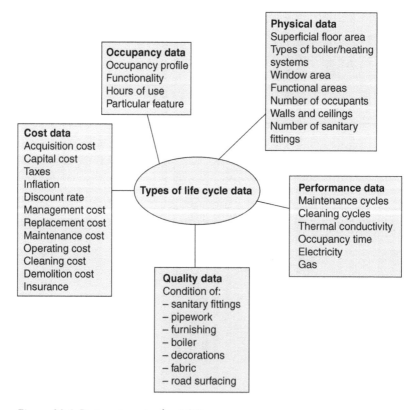

Figure 20.1 Data categories for LCC.

to be seen in the context of other categories in order to obtain the correct interpretation of them (Kishk *et al.*, 2003).

It should be noted that LCC is a decision-making tool in the sense that it can be used to select among alternative projects, designs or individual building components. Consequently, LCC data should be presented in a way that enables such comparison. For that reason, the cost breakdown structure is an important concept for LCC (Bakis *et al.*, 2003). There are several different standards (ISO 15686–5/NS3454/ASTM/Australian/New Zealand) available to guide a LCC analysis. All have different cost categories and cost breakdown structures.

Main sources of data

There are three main sources of data for LCC purposes:

1   from manufacturers, suppliers, contractors and testing specialists;
2   historical data; and
3   data from modeling techniques.

Data from manufacturers, suppliers, contractors and testing specialists can often be seen as a best guess. They may have a detailed knowledge of the performance and characteristics of their material and components, but do not have knowledge of the ways in which facilities are used (Flanagan and Jewell, 2005). However, extensive knowledge and experience of specialist manufacturers and suppliers are a valuable source for life cycle information. If the required data are not available, modeling techniques can be used. Mathematical models can be developed for analyzing costs. Statistical techniques can be incorporated to address the uncertainties (Flanagan and Jewell, 2005). Data from existing buildings are used as historical data, some of which are published by the BCIS in the UK.[3]

The quality of decision-making derived from the use of LCC calculations is constrained by the availability of appropriate and accurate data, which Flanagan and Jewell (2005) refer to as the 'data problem'. Thus, data collection brings difficulties; yet, LCC analysis is only accurate if the collected data are reliable (Emblemsvåg, 2003). Existing databases have their limitations; they do not record all necessary contextual information about the data being added to them (Kishk *et al.*, 2003). Data are usually expressed as units of cost which limits them to local use.

Integrated life cycle design

Apart from the difficulties of collecting data, the transition from understanding the theory of life cycle costing to practicing it is not easy (Flanagan *et al.*, 1989). In many cases, the intangibles (such as aesthetics) are in conflict with the results from LCC calculations (Kishk *et al.*, 2003), also contributing to the difficulties facing the usefulness of the technique. To conjoin these objective techniques with those of a more subjective nature, Flanagan *et al.* (1989) suggested using weighted evaluation matrices to handle the intangible costs and benefits. Notwithstanding, Öberg (2005) states that the majority of tools, such as LCC, durability of materials and environmental assessment, are limited to their specific purpose and cannot provide a holistic appreciation of the issue. A general and holistic model combining different tools for an optimal life cycle design has been termed integrated life cycle design (Sarja, 2002). The model, and the methodology linked to it, allows the multiple needs desired by owners, users and society to be handled in an optimized way across the total life cycle of the building. The main aspects included in the model are show in Figure 20.2. The model might be able to address several facets of uncertainty, going beyond a traditional LCC model. Notably, and importantly so, is that LCC calculations are included in this model.

Uncertainty and bounded rationality

Uncertainty is viewed for the purpose of this chapter as 'a business risk which cannot be measured and whose outcome cannot be predicted or insured against'. Two central contributors to uncertainty in a product development context are technology novelty/complexity and project complexity (Tatikonda and Rosenthal, 2000). Project complexity increases the degree of uncertainty as would occur in a construction project where, for example, a new frame material, new actors or a new type of cooperation was involved. Technology novelty is, in a product development context, defined as 'the newness, to the development organization, of the technologies employed' (Tatikonda and Rosenthal, 2000). This opens up the definition of technological novelty/complexity in construction to a broad range of attributes and possibilities including industrialized timber-framed housing. The continued development of this technology depends on

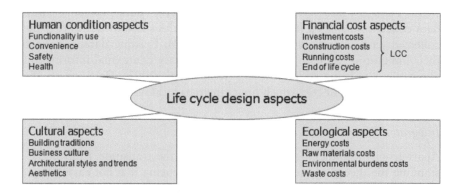

Figure 20.2 Main aspects of integrated life cycle design (after Sarja, 2002).

knowledge about its production process and product design being available to potential adopters so that their assumption of risk-taking decreases with increases in knowledge (Frambach, 1993).

Another central concept is that of the transaction, which 'occurs when a good or service is transferred across a technologically separate interface' (Williamson, 1985a). Two human and three environmental factors lead to transaction costs. The human (or behavioral) factors are:

1   *Bounded rationality.* Humans are unlikely to have the abilities or resources to consider every state-contingent outcome associated with a transaction.
2   *Opportunism.* Humans will act to further their own self-interests.

The environmental factors are:

- *Uncertainty.* Uncertainty aggravates the problems that arise because of bounded rationality and opportunism.
- *Small numbers trading.* If a small number of players exist in a market, there is little or no possibility of withdrawal and use of alternative players.
- *Asset specificity.* The value of an asset may be attached to a particular transaction that it supports. The possibility (threat) of a party acting opportunistically leads to the so-called 'hold-up' problem, which refers to the extent to which a party is tied in a business relationship.

It is not until the human factors are combined with the environmental factors that problems arise (Williamson, 1975). Bounded rationality is a problem only when it is combined with situations perceived as uncertain or complex for the party involved. As asset specificity and uncertainty increase, the risk of opportunism increases. Furthermore, human inclination for opportunism increases in a market with a small number of players, since opportunism brings its own punishment in a market with a large number of players (Williamson, 1975).

Transactions between organizations are controlled through contracts. This is especially true in construction since control of projects is based on firm contractual arrangements. However, bounded rationality makes it impossible to regulate every matter within a contract, a phenomenon Williamson (1985b) describes as contracts being unavoidably incomplete. In the relation between supplier and customer, trust is essential for cooperation between the parties, with contracts functioning both as a substitute for, and complement to, trust (Klein Woolthuis *et al.*, 2005). Prior experiences also play an important role in determining if and to what extent a partner can be trusted. Without long-term experience it may not even be possible to submit to the risk that the one who is trusted may actually fail (Nooteboom, 2002).

Research methodology

First, this section examines the different data that are needed to analyze LCC for buildings. The objective is to identify the main data required to carry out a LCC analysis and to move away from the limited application of LCC to a position where it can properly inform early design decision-making.

The incentive for an interview study, motivated in part by the earlier study of Höök (2005), was to deepen understanding of the uncertainties and lack of trust associated with timber-framed housing as expressed by building owners. According to a market analysis presented by a Swedish commission, *Industrifakta*,[4] clients are expected to gain increased power and to occupy a different role in future industrialized building processes. A key factor for clients is knowledge regarding the long-term performance of building systems. With a reflective understanding of the rationale for the uncertainties and lack of trust or confidence, the authors wished to investigate if and to what extent the uncertainties could be addressed by LCC considerations. With this in mind, we formulated the following research questions:

1   What are the different data needed to analyze LCC for buildings?
2   What are the preferable methods for LCC analyses for the construction industry?
3   What is needed to increase the use of LCC calculations and application in the early design stage?
4   What are the uncertainties expressed by building owners related to timber-framed housing and what is the rationale?
5   If and to what extent can aspects of integrated life cycle design be used to address the uncertainties?

Research methodology: interview study

The investigation was designed as an exploratory study with a qualitative research approach. Semi-structured, in-depth interviews were chosen as the main data collection method as they are especially suitable when the problem is complex. The interviewer can rephrase questions as well as pose related questions to penetrate the problem (Wiedersheim-Paul and Eriksson, 1989). The possibility to adjust questions for each individual was important in this case in order to get to the heart of the matter. The same questions, but from different angles, were asked thereby achieving a form of triangulation. In all, seven in-depth interviews with five building owners (of rental apartment buildings) were conducted. The participating owners were selected according to geographical location, size, type of ownership and whether they had experience of timber-framed housing or not (Table 20.2).

Table 20.2 Characteristics of the participating building owners

Location	Nr. of rented apartments	Ownership	Experience of timber framed housing
Luleå/Piteå	1,100	Privately owned	Yes
Stockholm	5,000	Cooperative economic association	No
Luleå	11,500	Publicly owned	No
All over Sweden	29,000	Privately owned	Yes
Stockholm	43,000	Publicly owned	Yes

The interview data were analyzed through categorization, taking into consideration the instructions of Dey (1993). An affinity diagram (Foster, 2004) was used for grouping the data against the aspects identified by Sarja (2002).

Research results

Results and analysis of the interview study

In analyzing the interview data, a number of clusters of uncertainties were distinguished. Sorting the different groups of uncertainties under headings aligned with the main aspects in *integrated life cycle design* resulted in a number of groups falling outside the model proposed by Sarja (2002). These could, instead, be clustered under the heading of *technical solutions*, identified by Höök (2005). The result of the categorization is given in Figure 20.3.

The figure shows the groups sorted under five headings and divided into three levels. The groups marked with an asterisk are uncertainties, or concerns, expressed especially about timber-framed housing. The groups in black squares are of highest importance as they are mentioned by four or more respondents. The groups on the second level, with gray background, were mentioned by three or less and the third (no background color) by one or two of the respondents. Of note is that no uncertainties were mentioned by four or more respondents under the headings of *culture* and *ecology*.

The most frequently mentioned uncertainties regarding *financial costs* are energy consumption, long-term performance and water damage. The respondents expressed a belief in higher energy consumption with a timber framed than for a traditional (concrete) frame and questioned the building's physical life and concerns about the consequences of water leakage in a timber-framed house. All building owners, except one, expressed concern about motion in timber frames causing, for example, cracks in wall finishes. All respondents expressed doubts that sound insulation, found under

Financial costs	Technical solutions	Human conditions	Culture	Ecology
Energy consumption*	Motions*/Stability*	Sound insulation*/ perceived sound level*	Cooperation with partners*	Natural and sustainable materials
Long-term performance*	Risk of fire*	Security/ safety	Experience, history, tradition*	
Water damage, piping, installations*	Fulfilment of functional demands on actual location*	Comfort, well-being	Dry building process*	
Maintenance of timber facades*	Timber as frame material*	Architecture, aesthetics		
Initial construction cost	Adaptability to new regulations and change			
Management and life cycle economy*				
Stairwells and timber staircases*				
Maintenance of facades, roofs and windows				
Serviceability, accessability				
Wear				

Figure 20.3 Uncertainties grouped and categorized.

human conditions, was good enough in timber-framed housing. Their impression was that the living environment would be disturbed by noise, making the building less attractive.

Most of the uncertainties are about cost in one sense or another. This finding clearly, and maybe not surprisingly, indicates that long-term cost is the most pressing uncertainty to address. Furthermore, a salient observation from the interviews is that cost is the decisive factor in design decisions, but with shifting focus on to short-term and long-term costs among the respondents. A difference could be discerned between private and public building owners. In summary, many of the uncertainties of highest importance can be addressed by LCC calculations, but not all.

A conclusion drawn from Table 20.3 is that the main grounds for the uncertainties about *technical solutions* originate from TVE housing being a new product offer, a new frame material, with a novel construction method (industrialized production), all of which can be referred to as *technology novelty*. The technology novelty together with high project complexity and economic value, and the need to incorporate a high degree of asset specificity mean that it is inevitable bounded rationality will influence the client's understanding of the transaction. Uncertainties concerning the technical solutions must therefore be addressed for the product to be trusted.

A building is a complex product delivered well after the contract has been signed. Human bounded rationality is, therefore, high for this type of transaction making trust an essential ingredient in the choice of contractor. No prior experience on the part of the contractor will lead to an even higher perception of risk-taking by the client. Furthermore, if the client has no prior experience of either industrialized production or timber-framed housing, there will be poor trust for all three inherent components in the choice of timber-framed housing and thus little motivation to take the risk. The small number of manufacturers in the timber-framed housing market further increases the perception of risk-taking. This can be explained by the increased inclination for opportunistic behavior by contractors, inevitably making them seem less trustworthy. Consequently, there is distrust in the product and uncertainty about the delivery, since the client has no possibility to withdrawal when there are no alternative contractors to engage.

One can derive many of the grounds for uncertainty in timber-framed housing, especially under the headings *financial costs* and *culture* from Sarja (2002) and *technical solutions* coined by Höök (2005). What is clear, however, is that most of the uncertainties can be embodied in the main aspects of integrated life cycle design (Sarja, 2002).

Conclusions

The choice of the appropriate method for LCC is easy and obvious if the advantages and disadvantages are appreciated. The calculation of LCC is not difficult and for structuring the main data, which need to be collected, help is available in the form of different standards (e.g. ISO/DIS15686–5 and NS3454). Nonetheless, data collection causes difficulty. Data need to be predictable if LCC analysis is to be reliable. Regional databases are seldom available or usable. Collecting data by hand takes much time and money. This may be worthwhile if the project is large enough. When historical data are collected and updated over time, their use can become more reliable and the LCC analysis more informative and valuable.

Data should be shared to avoid the duplicated effort of their collection. If more clients demand LCC information and a proper check of the information against performance is undertaken, improvement in accuracy and reliability could be expected. When LCC is used more frequently, the client would be able to judge LCC in the same manner as they do now with estimated capital costs. The client and end-users could save significant money in the long run if LCC is adopted as a decision-making tool. The lifetime quality and the cost effectiveness of buildings would improve also by using LCC in the early stages of design.

Timber-framed housing, as a comparatively new product on the construction market and produced by a new production process by relatively small and unknown manufacturers, generates uncertainties and skepticism among potential clients and building owners. The grounds for the uncertainties about timber-framed housing are shown to a large extent to be found in transaction cost theory and in the notions of technology novelty and project complexity. Knowledge of the rationale behind the perceived uncertainties makes it possible to address them. The study does not however reveal how this should be achieved since it was outside the scope of the current research.

Uncertainties relating to costs of one kind or another constitute the great majority of the uncertainties expressed by clients. One conclusion is that LCC calculations are able to address a large number of the perceived uncertainties in regard to timber-framed housing. The model must, however, be broadened to include all aspects in integrated life cycle design for the calculations to be applicable and acceptable. Furthermore, the characterstics of timber-framed housing are such that it crucial to address the uncertainties about its technical merits for it to be trusted by clients.

Acknowledgments

The authors gratefully acknowledge the financial support of *InPro*, an integrated project co-funded by the European Commission within the Sixth Framework Programme and *Trä Centrum Norr*, an R&D center at Luleå University of Technology. We also thank those who kindly gave their time for the interviews and helped to provide valuable information with which to test our ideas.

Notes

1 Buildings and constructed assets – Service life planning. Part 5: Whole life cycle costing.
2 Life cycle costs for building and civil engineering work – principles and classification.
3 www.bcis.co.uk.
4 Konsekvenser av industrialiserat byggande [Consequences of industrialized building].

21 Decision-making practice in the real estate development sector

Ellen Gehner and Hans de Jonge

Introduction

Managing a real estate development project is about risk-taking. From previous research (Akintoye and MacLeod, 1997; Lyons and Skitmore, 2004; Gehner *et al.*, 2006), it can be concluded that techniques of risk analysis are little used in the real estate development sector for reasons that include the paucity of objective data, time constraints and the lack of confidence in outcomes. Other methods are apparently used with success to manage risks in a project. A technocratic approach alone to risk management seems, however, insufficient to gain insight into these methods. The approach views risk as the probability of an event multiplied by the magnitude of loss related to that event (Raftery, 1994).

Risk should be regarded in a broader, cognitive and sociological context in which individuals perceive risk differently and their willingness to take risk varies (Sitkin and Pablo, 1992). Thus, risk must be seen in the context of choice (Stallen, 2002) or, in other words, 'risk is a consequence of a decision' (Luhmann, 1993). This implies that the way real estate developers treat risk is expressed to a large extent in the decision-making process. Much research has been carried out on decision-making, but not much in this sector and not specifically from this perspective. Research into the decision-making process is, therefore, needed in order to understand how risks are treated and how the risk management process could be facilitated in the real estate development sector. In this chapter, a single case study of a Dutch real estate developer is analyzed using a framework derived from the strategic decision-making literature in order to gain understanding of how the decision-making process is organized and contributes to risk management.

Literature review

During a real estate development process hundreds of decisions are made before a project comes to fruition. One group of decisions, so-called investment decisions, is a strategic action from the perspective of the developer and is examined here. The characteristics of the strategic decision-making process are reviewed and related to the stages of decision-making and risk management. These reviews result in an analysis framework for decision-making practice in real estate development.

Investment decisions in real estate development

Real estate development is a complex and lengthy process in which a wide range of risks is encountered. During the process, a developer is responsible for all activities that range from the purchase of land, coordination of the design and construction process, application for building permits and, ultimately, the sale of the resulting real estate (Peiser and Frej, 2003). The start of the process is characterized by a high degree of uncertainty, while initial investments are low. As progress is made uncertainty is reduced, while investment increases significantly in line with the signing of major contracts (Gehner and de Jonge, 2005). By signing a contract, the organization commits to a certain course of action and a corresponding investment or income. In taking this decision, the organization agrees to accept the risks that come with the commitments made as well as the extent of influence remaining for managing those risks. The major contracts in the development process are land purchase, cooperation agreement with the local authority (municipality), award of the construction contract(s) and real estate contracts (rental or sale). The signing of these contracts usually signifies a new phase in the process. At the gates between two consecutive phases, so-called investment decisions are made based upon contractual commitments.

Investment decisions are considered strategic for a number of reasons. In the first place, a decision is strategic when the decision is 'important, in terms of the actions taken, the resources committed, or the precedents set' (Mintzberg *et al.*, 1976) or, put another way, when it is one 'that critically affects organizational health and survival' (Eisenhardt and Zbaracki, 1992). From the perspective of a project, the importance of the decision is obvious; from the perspective of the organization the importance is high, as the financial resources committed to this kind of decision demand a considerable amount of equity. Depending on the nature of the contractual commitments, the same equity could be at risk in the worst-case scenario.

A strategic decision is also about 'the determination of the basic long-term goals and objectives of an enterprise' (Ghemawat, 1999). Although the investment decision is directed in the first place toward the course of action necessary for the project, it also directs the goals and objectives of the organization. The commitments made in a real estate development project are typically for a period of five to ten years. During this period the equity demand of the project restrains the organization from initiating new projects; thus, the mid–long term strategy is set by an investment decision.

A decision is strategic when it concerns 'those infrequent decisions made by the top leaders of an organization' (Eisenhardt and Zbaracki, 1992). Investment decisions are not made by the project manager, but only 'a person in authority' (Nutt, 2005), that is, someone at the strategic level of an organization is empowered to make such a decision. This analysis leads to the following definition:

> an investment decision is a commitment to the allocation of financial resources to a project taking into consideration the goals of the project as well as the objectives of the organization, with the decision taken at the strategic level since the risks involved could have significant adverse impact on the survival of the organization.

Two further comments must be made. First, strategic decision-making normally focuses on infrequent decisions; however, in each real estate development project several investment decisions are made, although do not differ so much in terms of the type of decision problem. Nonetheless, unfamiliarity with the decision problem remains, as each project differs largely in a spatial and economic context, parties involved and the program of requirements. Second, Dutch real estate developers are relatively small with simple organizational structures. Communication lines between the operational level of the project manager and the strategic level of decision-makers are shorter than is usually the case in examples described in the literature on strategic decision-making (see, for example, Mintzberg *et al.*, 1976; Eisenhardt and Bourgeois III, 1988; Nutt, 2004).

Strategic decision-making process

Decades of research into strategic decision-making has shown that these processes can be described in similar general stages such as *identification–development–selection* (Mintzberg *et al.*, 1976), *intelligence–design–choice* (Simon, 1977), *awareness–analysis–action* (Noorderhaven, 1995) and *framing–gathering intelligence–coming to conclusions–learning from feedback* (Russo and Schoemaker, 2002). These stages have similarities with the stages in risk management, being objective setting, risk identification and assessment, risk response, and risk control and monitoring (APM, 2004). Thus, looking at the stages of decision-making gives us at the same time insights into how risk management is carried out. The trouble with these models of decision-making is that the stages are too generic to serve as a framework for analyzing decision-making processes. A framework for analysis needs to consist of distinguishing factors regarding both the content of the decision and the subjects carrying out the process. Fredrickson (1983) identified six major characteristics to distinguish different types of processes, being 'process initiation, role of goals, the means/ends relationship, the explanation of strategic action, the comprehensiveness of decision making, and the comprehensiveness in integrating decisions'.

First, the process of initiation is related to the stage of identification and can be described in terms of *how* (reactive or proactive to a problem or risks) and *where* (at what level in the organization) the decision-making process is initiated. Second, goals play an important role in the selection stage: are they conceptualized in precise or general decision criteria, and are these criteria determined on the basis of individual or organizational objectives? A third question is whether the goals are irreversible and persistent regardless of the available means. This question is addressed in the development as well as in the selection stage. Both stages of development and selection can, as a fourth aspect, be explained by the extent to which a process is intentionally rational, the result of a social process or highly influenced by politics and power (Eisenhardt and Zbaracki, 1992). Fifth, the cognitive perspective is used to determine the comprehensiveness of each stage in the decision-making process: to what extent is a decision-maker influenced by cognitive limitations and perceptions? These questions all refer to an individual decision. The last characteristic is about the degree of integration between individual decisions within an overall strategy.

Decision-making and risk behavior

The six strategic decision process characteristics are used as a framework to describe decision-making practice in the case study organization (Figure 21.1). An analysis was undertaken of the extent to which a particular characteristic contributes specifically to the management of risks. The ultimate aim of the research – although not presented here – is to explain the mechanisms that steer the decision-making process and, thus, risk behavior. Individual characteristics influencing risk perception and propensity (March and Shapira, 1987) are taken into account as well as organizational factors. Fredrickson (1986) explored the relationship between the structure of an organization and its strategic decision-making process and suggested that cultural aspects, as well as financial structure and point in the life cycle of the organization, influence risk behavior.

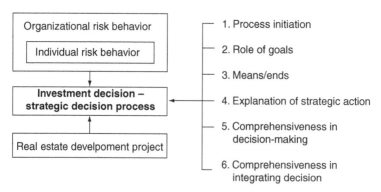

Figure 21.1 Case study analysis framework.

Research project

Here, the preliminary results from the study of a Dutch real estate developer are presented. The developer had a portfolio of around €1.0–1.5 billion (in 2006) (van Enk, 2006), reflecting a mix of retail, leisure, commercial and business-to-business use. This case study is part of wider research (Gehner, 2008) in which a multiple case study is being carried out to explore the strategic decision-making processes in the real estate development sector. The aim of that larger study is to analyze risk management strategies within the development process and to explain differences in risk-taking behavior from the dual perspective of the organization and the individual.

Project description and objectives

As noted earlier, investment decisions in real estate development can be regarded as strategic, reflecting the risk behavior of the developer. As the application of formal risk analysis techniques is often not used by real estate developers, the aim of this research is to gain an understanding of what risk management strategies – other than or in addition to formal techniques of risk analysis – form part of the strategic decision-making process. In order to obtain insights, the following research questions were formulated.

1   How are investment decisions made in real estate development?
2   What risk management strategies are, explicitly or implicitly, incorporated in the decision-making process?
3   Which characteristics of the organization, as well as of the individual, are explanatory for the purpose of revealing risk behavior?

The strategic decision-making process makes use of the six characteristics described by Fredrickson (1983) (see Figure 21.1). The extent to which the process contributes to managing risk in a real estate project is questioned in the context of each of these characteristics. Finally, some hypotheses are presented that might explain the risk behavior of a real estate developer. Testing these hypotheses is the starting point for further research.

Research methodology

The objective of this research implies an in-depth, holistic analysis of the decision-making process of a real estate developer to explore its use, explicitly and implicitly, of risk management. In the study of a social phenomenon, in general, three research strategies can be distinguished: survey, experiment and case study (Braster, 2000). A survey is well suited to researching social phenomena on a large scale, but only superficially. In order to conduct a survey it is necessary to have a well-defined conceptual model with a limited number of variables. Quantitative methods are used to analyze data so the statistical validity of this research strategy is high, but it does not give profound insights into the phenomenon under study. An experiment is an appropriate method for gaining deeper knowledge by conducting a controlled test of a hypothesis. The general procedure is that one or more independent variables are manipulated to determine their effect on a dependent variable, while so-called antecedent or intervening variables are eliminated or controlled. Experiments are carried out by cognitive psychologists on individual decision-making in game situations and by social psychologists on group decision-making in the laboratory (Mintzberg *et al.*, 1976).

A case study is well suited to understanding the *how* and *why* of phenomena in their natural settings (Yin, 2003). Multiple research methods are used to collect data on a wide variety of variables. A qualitative method is used to analyze the data in contrast to both surveys and experiments where quantitative methods are more appropriate. We chose a case study because the nature of the research is exploratory and not limited to fixed variables and their causal relationships. We also wanted to research the decision-making process within the real life context of the organization. All variables are taken into consideration and not decoupled as in an experiment.

Case selection

The overall case study design consists of three cases for literal replication (Yin, 2003). This chapter presents the results of the first case study. For each study, a real estate developer was selected based on its portfolio, financial independency, track record/image and internal stability. We made a selection of Dutch real estate developers, with their development portfolios in the Netherlands, having an investment value over €250 million ((m^2 × rental value)/(m^2 × gross yield)). The reason for using this criterion is that such organizations are able to develop projects of a substantial size with corresponding risks. The second criterion is related to the different types of developers that are found in the Netherlands: those associated with a financial institution, contractor, investor, housing corporation and independent developers. We selected independent developers only as the core business of these organizations is to develop real estate projects for a financial return. That aim is not therefore impaired by the business objectives of other operational units. The third criterion is a good track record and entrepreneurial or risk-taking image within the construction industry. This is not a hard, objective criterion, but gives an idea of how an organization is perceived by other parties involved. Moreover, a good track record implies that risks are adequately managed. The final criterion is the internal stability of the organization. Structural change, such as a merger or acquisition, can strongly influence the decision-making process. For reason of confidentiality, the selected real estate developer is anonymous.

Data collection

Within the case study, multiple methods were used to collect data to counter the intrinsic subjectivity of the research strategy. In the first instance, a total of 13 in-depth interviews were conducted with 12 representatives of each part of the organization (Mintzberg, 1979): operational level (four project managers); middle management (one program manager); strategic level (three members of the board); techno-structure (two financial analysts) and support staff (two secretaries). The interviews varied in length from 25 to 90 minutes, with the average around one hour. We used semi-structured interviews based on a list of topics for which the interviewees did not have to prepare themselves. For each interviewee, the topics were adapted to his/her specific functional level in the development project and his/her role in the decision-making process. During the interviews, notes were taken and used to produce transcripts. Unless the interviewee objected, the interview was recorded and the recordings were used to supplement the transcripts.

As a complementary measure, documents covering the decision-making process – taken in the broadest sense – and the three projects were analyzed. They included strategy documents (e.g. organizational structure, annual report and code of conduct), policy documents (e.g. project handbook and meeting schedules) and operational documents (e.g. budget request form, budget reports and project reports).

The third method employed to collect data was participatory observation. Over the seven weeks in which the case study was conducted we attended all strategic meetings concerning investment decisions. In these meetings notes were taken on the issues discussed, the originator of those issues, how a conclusion was reached and how the decision was formulated.

Data analysis

The approach described here is directed at the single case study which served as the basis for cross-case analysis. The first step in the analysis was a description of the intended decision process derived from formal documents and partly from the interviews. The second step was the description of the actual decision process observed during the strategic meetings and derived from the interviews focusing on three projects. The third step was to deduce risk management strategies from the strategic decision process: how does the decision process contribute to the management of risks? The final step was the comparison of these two processes and searching for explanations in existing theory for differences between the actual and intended decision process and risk behavior.

Research results

The investment decision process is described following the characteristics presented in Figure 21.1. Each characteristic is used to indicate the extent to which the decision process contributes to risk management.

Results

Process initiation

In the organization's project handbook, the development process is divided into six phases: acquisition, initiation, pre-development, development, construction and operations (i.e. facility management). Between two consecutive phases an investment decision has to be made (Figure 21.2). The investment decision is concerned with allocating a budget to cover expenditure on a specified set of activities in the coming phase. According to the organizational structure, it is the responsibility of the project manager to initiate the decision process. A new budget has to be requested by the project manager before starting new activities and when the budget for an incomplete activity is likely to become overspent.

The phasing of the development process, including integral decision gates, and the allocation of a fixed budget to each phase – thus restricting decision authority – are the primary means for managing risk. Unfortunately, budgets and activities are coupled without taking the factor of time in account. This means that investments can be made and activities allowed to produce no result, yet go undetected. A monitoring system could be a solution to this problem, which takes away some responsibility from the project manager.

Role of goals

For each decision gate, explicit decision criteria are formally determined. These criteria are conceptualizations of organizational goals at the level of detail of activities in each phase of the project. For example, in the concept phase an inventory must be prepared on the (anticipated) zoning plan, a program of requirements has to be formulated and a market feasibility report is required. The criteria become more measurable in the course of the process; for example, in order to start construction, the main contract must be tendered, the building permit must have been obtained and over 70 percent of the rentable area must be leased. In contrast to goals relating to the status of the project at any particular decision gate, future goals are set in terms of the expected return on investment.

Figure 21.2 Phases and investment decisions in the real estate development process.

Determining the preliminary results for a phase and using them as the decision criteria for entering the next phase is a risk management strategy. This strategy achieves a certain amount of certainty by limiting the extent of risk to be taken in the next phase through agreement on future financial commitments in the form of budget allocations. This openness on goals gives the project manager freedom to operate within clear boundaries, whereas decision-makers are forced to act according to these goals, not their own. The downside to preset criteria based on profitability is that there is little room for other strategic goals to carry the decision.

Means/ends relationship

The goals mentioned in the previous section, the expected return and the preliminary results constitute the minimum criteria. An unconditional budget approval (the so-called 'go' decision) is possible only when all criteria are met. Deviations from these criteria are not acceptable, unless

1   budget and time are limited; or
2   conditions regarding the activities (i.e. an alternative strategy) are set.

These limitations are part of the decision and, essentially, measures for optimizing the development strategy (allocation of means) while, at the same time, managing risks by:

1   limiting expenditure; or
2   reducing or transferring risk by improving the deal or financial structure.

Decision-makers must not become inflexible because of set criteria, but should act as an advisory body for the project manager.

Explanation of strategic action

Since the decision criteria are not arbitrary, but explicit and familiar both to project managers and decision-makers, the decision-making process can be described as intentionally rational. Still, good information is necessary to make an adequate, rational choice. The project information is provided by the project manager, which is always influenced by subjective perception. To counter this weakness, the decision is formulated as a two-step process. First, the project is assessed on the progress of activities by program managers. If they approve the budget request, the project is evaluated on its financial merits by the investment committee.

The structure of the decision process functions as a risk management method in the sense that it strives for intersubjectivity. Dividing the decision into two parts based on different fields of interest and convening groups of experienced people with varying degrees of involvement in the project contribute to an objective, rational outcome.

Comprehensiveness of decision-making

The two-step decision-making and group composition already help to reduce cognitive limitations during the process of *intelligence–design–choice* (Simon, 1977). Yet, the extent to which decision-making can be regarded as comprehensive depends largely on the quality of the information provided by the project manager. Information accompanying the budget request consists of a memorandum and an investment figure. The memorandum is presented in a standardized format consisting of five categories of activities (corresponding with the decision criteria) – land, building/permits, budget, marketing and financing – in which the status is described. For the purpose of risk management, a qualitative description of risks is added, but no formal quantitative risk analysis is undertaken. A visual presentation of progress in the activities in the form of a spider diagram can be used to supplement the memorandum.

Decision-makers prepare by reading the budget request, arriving at their decision during a group discussion. During this discussion, additional information is sought, assumptions are questioned and alternative scenarios (risks) and strategies (risk measures) are (unsystematically) elicited and evaluated. Finally, a decision, possibly with additional conditions or limitations, is proposed and refined until unanimity is reached.

Comprehensiveness in integrating decisions

Standardization of goals implies equal treatment of all projects in the overall strategy of the organization. Even so, investment decisions need to be integrated with cash flow planning as the organization has a limited amount of equity at its disposal. Diversification in type and phases of projects is desirable so that cash flows are evenly distributed. Moreover, there is an attempt to keep initial investments low and finance substantial projects with outside capital from private equity providers or financial institutions, thereby transferring risk in exchange for a share in the return. The bottom line for allocating budgets to (new) projects is that losses from failure in all projects is less than the organization's equity capital, so the survival of the organization is not endangered.

The financial structuring of a project is an important method for managing risks, especially the risk of failure. Few people have insights into the total cash position of the organization and so only they are able to take account of this factor in decision-making. To enable others to make a comprehensive decision, a rule of thumb has been provided on the acceptable level of risk from project failures.

Implementation and exploitation

The lessons arising from the case study are concerned with the structure of the decision process, the kind of information used to support decision-making, ways in which information is handled and risk management strategies to be applied. As a next step, the findings should lead to recommendations on the efficiency and effectiveness of implementing decision support tools to better respond to, and control risks in, the real estate development process.

As shown in Figure 21.1, a decision not only depends on the risk profile of the project, but also on the characteristics of the organization and the (group of) individual decision-maker(s). In terms of the behavior of individuals, much research has been published in the field of cognitive psychology (see, for example, Tversky and Kahneman, 1974; March and Shapira, 1987; Sitkin and Pablo, 1992; Simon *et al.*, 2000). As to the influence of organizational structure, Fredrickson (1986) described the influence of three archetypical structures on the strategic decision-making process.

From this particular case study, two characteristics seem to influence risk behavior significantly. The first is the culture of the organization which, in this case, is strongly based on trust and entrepreneurship. Since entrepreneurship is delegated to the operational level, considerable responsibility is given to project managers, who act within a set of decision criteria. When making an investment proposal, the decision-makers rely on the honesty, integrity and knowledge of the project manager. The second characteristic is the equity position of the organization. In the case study, the financial structuring of a project was emphasized, since the equity position is limited and, therefore, risk-taking capacity is limited. In further research, more attention needs to be given to the influence of the organizational structure, culture and (financial) means on risk behavior.

Conclusions

Investment decisions are presented as strategic decisions within the real estate development sector. These decisions are about allocating financial resources to a development project and, as a consequence, accepting the risks that result from this commitment. As risk is a consequence of a decision, decision-making and risk management are intertwined. To get a better understanding of the way real estate developers treat risk, the decision process within a developer was analyzed in accordance with recognized strategic decision-making process characteristics.

From the analysis, it can be concluded that a series of risk management strategies is sometimes explicitly, but also implicitly, embodied in the decision-making process. In the first instance, the gates between two consecutive phases in the development process allow for a decision to be made regarding the allocation of a fixed budget to the coming phase, thus limiting the amount of risk that can be taken. To support this decision-making process, a set of decision criteria is explicitly determined that guarantees a particular level of certainty and, therefore, a level of risk reduction. This set of criteria serves as a guide against which one can only deviate if it is proposed to manage risk in a different way; for example, by transferring risk or taking more risk but limiting it to a certain point or to a fixed amount. To respond adequately to the investment proposal and related risk, one strives for intersubjectivity through use of a two-step decision process and by convening groups of experienced people drawn from different disciplines. During discussion leading to a decision, these decision-makers weigh several scenarios, ascertain that the worst-case scenario does not endanger the survival of the organization and develop fall-back strategies. Based on this unsystematic risk analysis they reach a decision.

This research has shown that the approach to examining risk management by scrutinizing the decision-making process is relevant. Although the results are based on one case study, the research will be extended through more empirical work. The final aim is to contribute to the implementation of efficient and effective risk management strategies. These will not be based on formal techniques of risk analysis, but on the conditional requirements of organizational risk behavior.

22 Differences in the application of risk management

Ekaterina Osipova and Kajsa Simu

Introduction

Construction projects are usually characterized by many and varying risks. Being able to manage risks across all phases in the construction process is an important and central element preventing unwanted consequences, not least exposure of the client to financial or other losses and quality failures. Many different actors are involved in a construction project and often they have no or limited experience of earlier collaboration with each other. It is not unusual for actors to avoid risks as far as possible and let somebody else in the value chain deal with them. The generally accepted principle is that, in each phase of a construction project, the management of a specific risk should be allocated to the party that has the best ability to handle it. In the context of small projects, where risk management might not be formalized as in larger projects, understanding of the risk management process in general and mitigation steps in particular might be wanting.

Many actors are involved in just some of a project's phases and this lack of continuity has been identified as one source of problems relating to quality and cost (Yngvesson *et al.*, 2000; Ericsson *et al.*, 2002). These actors tend to focus on short-term financial results and protect their own interests rather than worry about the project. Such behavior leads to less effective management of risks overall. From our preliminary investigations, little attention has been paid to identifying the roles of individual actors in risk management across the project's phases. No research seems to have been directed at small projects, i.e. those running for a matter of months with contract sums in the region of SEK1 million to SEK15 million,[1] typically ground works, minor building works and refurbishment. They are believed to represent in the region of 80 percent of all projects by number and whose aggregated value is significant at roughly 20 percent of total construction output (Hultén, 2004).

The objective of the chapter is to analyze the risk management process on construction projects from the perspective of the client, contractor and consultant in general and in the context of small projects in particular. We examine the ways and extent to which the main actors are involved in risk management across the different phases of projects and compare practices with the generally received view of risk management (Osipova and Apleberger, 2007). The particular case of small projects is given closer examination to understand the extent to which systematic risk management takes place and the methods and tools used in this work. Our research is based on literature reviews and the results of two surveys carried out in parallel.

Literature review

Risk and risk management in construction

There are several definitions of project risk in the literature (see, for example, IEC, 2001; PMI, 2004; Baloi and Price, 2003; Barber, 2005). A formal definition is given in the international standard IEC62198 as 'a combination of the probability of an event occurring and its consequences for project objectives'. Ward and Chapman (2003) discuss the concept of risk in greater detail and suggest using the more general concept of uncertainty. A questionnaire survey conducted by Akintoye and MacLeod (1997) shows that the majority of project actors perceive risk as a negative event.

Project risk management is a formal process directed to identification, assessment and response to project risks. The process is defined differently in research literature (see, for example, Flanagan and Norman, 1993; Uher and Toakley, 1999; Chapman and Ward, 2003). Even so, all definitions align with the aim of maximizing opportunities and minimizing the consequences of a risk event in the construction project. PMBOK® (PMI, 2004) identifies four main steps in the risk management process: risk identification, risk assessment, development of risk response and management of risk response. Several authors develop more detailed models; for instance, Baloi and Price (2003) use the model of seven steps: risk management planning, risk identification, risk assessment, risk analysis, risk response, risk monitoring and risk communication. Chapman and Ward (2003) introduce the SHAMPU model, which consists of nine phases. Del Cano and de la Cruz (2002) present a generic project risk management process of 11 phases, which can be used in large and complex projects. For the purpose of our research, we have adopted a simplified risk management process of three main steps: risk identification, risk assessment and risk response. The reason for the simplification is that this model is well known among project actors and commonly used in practice.

The goal of the risk identification process is to decide on potential risks that may affect the project. There are several approaches for classifying project risks and risk sources (Leung *et al.*, 1998; Tah and Carr, 2000; Baloi and Price, 2003; Li *et al.*, 2005). The main categories are financial, economic, managerial, legal, construction, design and environmental risks.

During risk assessment, the identified risks are evaluated and ranked. The goal is to prioritize risks for management. Baccarini and Archer (2001) describe a methodology for the risk ranking of projects, which allows for an effective and efficient allocation of the resources needed to manage risks. The JRAP model proposed by Öztas and Ökmen (2005) is a pessimistic risk analysis methodology, which is effective when encountering uncertain conditions in construction projects.

Several surveys conducted in construction (Akintoye and MacLeod, 1997; Uher and Toakley, 1999; Lyons and Skitmore, 2004; Edum-Fotwe and Azinim, 2006) show that checklists and brainstorming are commonly used techniques in risk identification; subjective judgment, intuition and experience are used mostly in risk assessment; and transfer, reduction and avoidance are the most applied methods in risk response.

Project phases in risk management

The construction process can be divided into four main phases: concept,[2] design, procurement and production. Since it is impossible to foresee all project risks in the concept phase where there is so much uncertainty, and because of the likelihood of identified risks changing during project implementation, joint and consistent risk management is required throughout all of the project's phases (Rahman and Kumaraswamy, 2004). Motawa *et al.* (2006) propose a model, which helps in determining potential changes in the project based on available information in the early stages. Baccarini and Archer (2001) introduce a methodology for a risk rating process in the procurement phase, which allows for the effective and efficient allocation of resources for project risk management.

Several authors have highlighted the importance of the early phases in project risk management since the decisions taken in these phases can have a significant impact on the final result (Kähkönen, 2001). According to Uher and Toakley (1999), the actual use of risk management techniques in the early phases is very low. More recently, Lyons and Skitmore (2004) investigated the use of risk management in each of the project phases. Their results showed that risk management in the design and production phases was higher than in the concept phase; whereas, risk identification and risk assessment were undertaken more often than risk response.

Research methodology

Study of the risk management process

The first study – the risk management process in general – involved nine construction projects (Table 22.1) which met the following requirements:

- the projects were located in both large and small cities;

- they used different forms of contract and collaboration, i.e. design-bid-build contracts, design-build contracts and partnering;
- the projects were a mix of building and civil engineering;
- all projects, except one, were classed as medium-sized (i.e. between SEK15 million and SEK100 million).

As the objective of the study was to obtain a picture of the risk management process from different actors' perspectives, a questionnaire survey was selected as the most appropriate research method. The survey sample comprised clients, contractors and consultants. Within each group, we identified those persons who worked with risk management on the project. The respondents from the client's side are the person signing the contract and project manager. From the contractor's side the respondents are the person signing the contract, site manager and cost estimator. Finally, the respondent from the consultant's side is the architect or design manager.

A draft questionnaire was developed consisting of five sections. The first section contained general questions about the respondent. In the second section, aspects of the risk management process across the different phases of the project were covered. The third section investigated relationships between the actors in the project. The fourth section focused on software management systems, which the company uses in its risk management process. The fifth section gathered miscellaneous comments regarding the risk management process on the project.

We organized two workshops where we met about 50 percent of the prospective respondents and presented the research project and objectives of the survey. The workshop participants were given an opportunity to answer the draft questionnaire and gave their comments on the content. Following the workshop, the final version of the questionnaire was developed and sent in electronic form to the respondents. After the questionnaires were completed and returned, the responses were analyzed using the statistical package, SPSS, and Microsoft Excel.

Table 22.1 Characteristics of projects included in the study

Nr.	Location	Type of project	Form of contract/ collaboration	Contract sum (SEK millions)
1	Norrbotten	Building	Design-build	41
2	Norrbotten	Building	Design-bid-build	18
3	Norrbotten	Other civil engineering	Design-build	53
4	Norrbotten	Road	Design-bid-build	20
5	Norrbotten	Road	Design-bid-build	5
6	Stockholm	Building	Design-build	81
7	Stockholm	Building	Design-build	48
8	Stockholm	Other civil engineering	Design-bid-build	95
9	Stockholm	Building	Partnering	15

Results of the survey

In total, 54 questionnaires were sent and 43 responses were received, resulting in a response rate of 80 percent. From the received responses, 36 were complete and seven respondents explained the reasons for their non-completion of the questionnaire. All those who attended the workshop responded. This would seem to indicate that respondents who were aware of the survey objectives were more interested in taking part in the project. The sample composition aggregated according to actors' roles in the project is shown in Figure 22.1.

Respondents. 34 survey participants were male and two participants were female. The age distribution reveals that 89 percent are over 41 years old. Most of the respondents (92 percent) have more than ten years' experience in the construction industry and 64 percent have more than 20 years' experience. Forty-four percent of survey respondents have a university degree in construction, 53 percent finished upper secondary school and just one person received vocational training. Thirty-three percent of the respondents say they participated in risk management or project management courses within their organizations or during a period of formal education.

Despite a relatively high education level and large experience, the majority of the respondents (75 percent) rate their knowledge of risk management as fair. Table 22.2 summarizes the risk management knowledge within each group.

Risk management in the different phases. The majority of respondents (32) participated in the production phase (Figure 22.2). For contractors, this is rather expected and obvious. Surprisingly, just seven clients participated in the concept phase compared to 14 in the production phase. This may be partially explained by the types of the projects. Often, there is no concept phase as such in civil engineering projects. All four consultants participated in the design phase and two of them followed into production.

When respondents were asked to rate[3] the importance of risk management in each phase of the project (Figure 22.3), the results were similar for both the client and contractor groups. The production and design phases

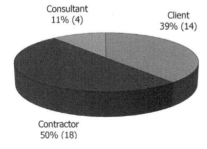

Figure 22.1 Sample composition.

Table 22.2 Knowledge of risk management

	Role in the project			Totals*
	Client	*Contractor*	*Consultant*	
Low	1	0	1	2
Fair	10	14	3	27
Advanced	2	3	0	5

Note
*Two respondents did not answer this question.

were identified as the most important for managing risks: the procurement and concept phases then follow. Consultants' ratings differ from those of clients and contractors. Overall, we can see that they underestimate the importance of risk management all phases compared with the rest. Not surprisingly, the design and production phases are identified by consultants as the most important for risk management. From this distribution, we can conclude that many actors associate risks with the production phase.

Those carrying out risk management processes systematically in their projects can be seen in Figure 22.4. The most active group is contractors, where all respondents identified and assessed project risks and 94 percent performed risk response systematically. In the client group 86 percent identified risks, 71 percent assessed them, while 57 percent systematically responded to project risks. The explanation for the low risk response rate may be that the clients let other actors in the value chain deal with

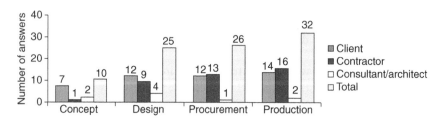

Figure 22.2 Participation by project phases.

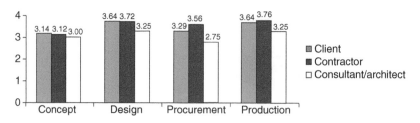

Figure 22.3 Perceived importance of risk management in the different phases.

identified risks. Consultants are the most passive actors when it comes to project risk management. Among consultants only 33 percent identified risks and responded systematically, and none assessed project risks.

Risk identification process. Risk identification was mostly performed in the design and production phases (Figure 22.5). The earlier that risks are identified, the less is the probability they will occur. Yet, just seven respondents indicated that risk identification was performed in the concept phase. Most of clients indicated that risk identification was carried out in the design phase, whereas contractors mostly identify risks in the production phase.

In the concept phase, 75 percent of the respondents answered that risks were identified by the client. This can be contrasted to the design phase where 39 percent responded that risk identification was performed jointly by all actors and 25 percent responded that it was performed by the client and the consultant. In the procurement phase, the contractor plays the most important role in risk identification (52 percent). In the production phase risks were identified by the contractor (39 percent) or jointly by all actors (39 percent).

Risk assessment process. Figure 22.6 shows that risk assessment has a similar tendency as the risk identification process: the majority of respondents undertake it in the production phase. The procurement phase is, however, more important for the risk assessment process than for risk identification and risk response. This is because the risk premium is calculated in the procurement phase and therefore it is important to assess earlier identified risks.

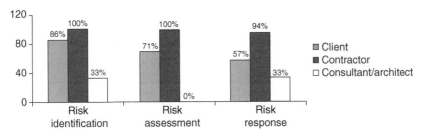

Figure 22.4 The risk management process systematically performed in the project.

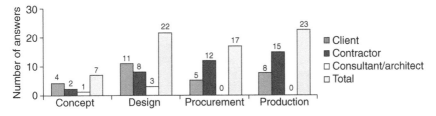

Figure 22.5 Risk identification in the phases.

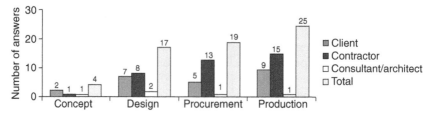

Figure 22.6 Risk assessment in the different phases.

Similarly to the risk identification process, risk assessment in the concept phase is performed mostly by the client, in the design phase jointly by all actors or by the client and consultant. The contractor's involvement in risk assessment in the design phase was higher than for risk identification. The procurement and production phases do not differ much from the risk identification process: in both phases the contractor plays the most important role.

Risk response process. Risk response is also associated with the production phase (Figure 22.7). Clients and contractors mostly manage risks in this phase, reflecting the longstanding custom in the industry where contractors tend not to put so much effort into preventing problems, preferring instead to solve them as they arise.

In the concept phase, risk response is performed by the client in line with our findings for risk identification and assessment. In the design phase the client, together with the consultant, responds to the project risks. In the procurement phase risk response is performed mainly by the contractor. In the production phase, the role of the contractor is large and the degree of joint risk management is high.

Collaboration in risk management. In the questionnaire, we defined the term collaboration as joint work in the risk management process. Almost all respondents had collaborated in risk management with other actors in the project: 11 clients, 13 contractors and three consultants. Seven respondents (three clients, three contractors and one consultant) said that no collaboration in risk management existed in the project. The rating[4] of collaboration (Table 22.3) varies from 'fairly good' to 'very good'.

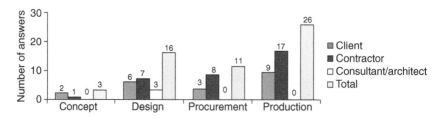

Figure 22.7 Risk response in the different phases.

Table 22.3 Evaluation of collaboration in risk management

Role in project	Evaluation
Client	3.55
Contractor	3.38
Consultant/Architect	3.33

The extent[5] of communication of known risks and opportunities between actors in the procurement phase is presented in Table 22.4. Overall evaluations are not high and vary between 'little detailed' and 'fairly detailed'. The contractors indicated that the client communicated known risks moderately (2.06). On the other hand, the clients indicated that their communication of known risks was higher (2.73).

Figure 22.8 presents the respondents' judgment[6] of their own and other actors' influence on risk management in the project. The results show that the contractor has the largest influence on risk management from the perspective of all actors. It is interesting that even clients estimate the contractors' influence to be larger than their own. This can be linked to Figure 22.3, where the actors associate risk management with the production phase. The influence of the consultant/architect is surprisingly low despite the design phase being considered very important by all actors.

Collaboration in risk identification, risk assessment and risk response is shown in Figure 22.9. Risk identification (RI) is the process where collaboration existed according to most of the actors: 82 percent of clients, 92 percent of contractors and 67 percent of consultants indicated that they collaborated in identifying the project's risks. During the risk assessment process (RA), both clients and contractors collaborated with each other, while only 33 percent of consultants indicated that collaboration existed. The risk response process (RR) sees a lower degree of collaboration according to contractors: 62 percent of them had collaborated in managing risks.

The existence of collaboration across different phases is presented in Figure 22.10 and reveals a low level of collaboration in the concept phase. Just 14 percent of clients, the most active participants of the concept phase,

Table 22.4 Degree of communication of known risks and opportunities between actors in the procurement phase

	Clients' communication	Contractors' communication
Client	2.73	2.69
Contractor	2.06	2.39
Consultant/Architect	3.00	3.00
Total	2.36	2.53

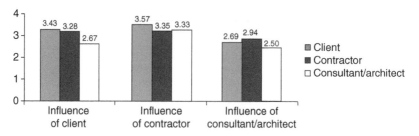

Figure 22.8 Influence of the actors on the risk management process.

said that collaboration existed. In the design phase, 70 percent of clients, 75 percent of contractors and 100 percent of consultants collaborated in risk management. This result can be linked to the importance of risk management in that phase, which was ranked high by the actors. In the procurement phase, collaboration between the clients and the contractors in risk management existed on half of the projects. In the production phase, collaboration between the actors is the most intensive, because many risks appear in this phase and have to be managed.

Study of risk management on small projects

Most studies into risk management in construction tend to cover large-scale projects, usually involving a multiplicity of participants (Ahlenius, 1999; Hintze, 1994; Jaafari, 2001; Hintze *et al.*, 2003). The risks cover a spectrum of events from financial, political and legal to technical and managerial, often related to complex construction. Far less is known about risk management on small projects, especially at the site level. The second study presented in this chapter focuses on small construction projects and the

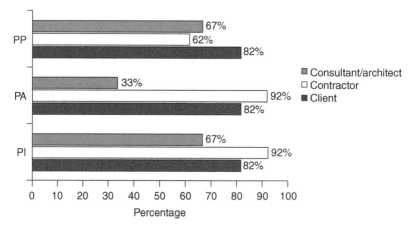

Figure 22.9 Existence of collaboration in the risk management process.

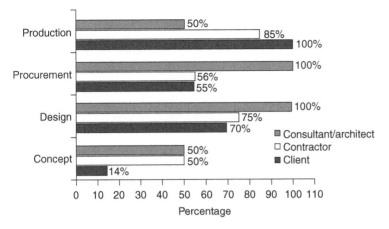

Figure 22.10 Existence of collaboration in the project's phases.

associated approach to risk management. The aim was to determine the nature and extent of risk management on small projects and to compare that with the received view obtained from the literature.

The characteristics of a small-sized project were discussed and agreed with the study's reference group.[7] The resulting description is enough to distinguish these projects from very small projects, which are more in the nature of continuous business operations, and from large projects. The agreed characteristics were:

- contract value between SEK1 million to SEK15 million;
- a site manager responsible for a maximum of two projects simultaneously;
- relatively short construction period (maximum 12 months);
- established techniques (no development work);
- project environment is independent; and
- personnel involved are more generalist than specialist.

Projects under SEK1 million were excluded because they were considered to offer very few insights from a project risk management perspective. The upper limit at SEK15 million was chosen in line with other parameters for a small-sized project.

The study was focused on the site level and took the form of an interview-based survey, benefiting from lessons drawn from case study strategy (Yin, 2003). The decision to employ an interview survey, in which qualitative data would be generated, was both strategic and pragmatic. Knowledge of risk management and its terminology is limited on construction projects, especially at the site level. Dialog was felt to be preferable to the completion of a questionnaire as the former would help to avoid misinterpretation and misunderstanding. Semi-structured interviews enabled respondents to add information and the researcher was free to adapt questions when a promising line of inquiry ensued.

The inevitable question arose about the appropriateness of projects for the study and their number. With the assistance of the reference group, ten projects were considered sufficient for our purpose. Interviews were held with three individuals having different roles in each project. The key role was that of the site manager, then the company's project manager and client's project manager. Data collection would thus be based on about 30 interviews. In the end, 28 interviews were conducted, since two project managers were responsible for two projects apiece. Documentation about each project was provided and analyzed.

Results of the survey

There are some very clear results from this study, one of which was that most managers rely on their intuition and conventional ways of controlling the project rather than on any formal risk management system. The largest group of respondents (17 out of 28) applied a similar approach to risk assessment and are especially interesting, not least because they represent such a large proportion of the total group. The following two quotations support their approach: 'The actual assessment is based on experience and very much in my head'; 'These risk checklists are filled in, but the meetings and the shared thoughts happen during coffee breaks or over the phone'. In a sense, this is the traditional way to control projects in construction. It starts with the unstructured allocation of 'risk money' in the bid. Money is allocated on the basis of instinct and experience, rather than in any systematic or formalized manner. Two of the respondents put it this way:

> I do not document the risk assessment I do; I put in some extra money in the bid. Sometimes I allocate money specifically for a certain risk, but not so often.

> Sometimes we allocate money in the bid for the site manager to see. The money is a combination of that allocated to prevent failures and as a measure to handle things that go wrong; but it is mainly for measures if things go wrong.

Controlling a project relies heavily on resource-based schedules, which is a matter of good practice. Critical events are documented as part of quality management, if they are documented at all. A common way of controlling issues in the quality assurance plan is to produce detailed work plans, either verbally or in writing. Meetings are the places where time, money and contractual conditions are handled. These features are found in most of the projects covered by this study. These steps are not, however, necessarily kept together and a formalized approach is lacking.

Another finding is that of a lack of training and education of risk management across the projects. More than half of the respondents had not received any education or training in risk management and the rest received it as a minor part of something else.

The role of the client in the project risk management is also rather passive and even strange. Clients place demands on the contractor to undertake risk management, but do not contribute to the process. Moreover, they expect the contractor to do so with the contract documents and nothing else. Information about the project prior to construction is not shared with the contractor. One of the clients described it this way:

> We have requirements in our management system to manage project risks, but so far we have only stated that the contractors should do this. This is a new way of working and so far it has only been applied in one project and not in the one we are talking about.

Another client said almost the same thing about its way of handling project risks: 'We do require that the contractors have management systems for quality and environment, but we do not have our own system in use. We are working on the development of such systems'. Risk identification and risk assessment are formalized on very few projects. Most often these tasks are undertaken on plain paper based on the individual's experience and personal judgment. The use of risk management systems is neither obvious nor a matter of course. Instead of using a management system and risk tools, site managers control their projects almost entirely through schedules. Planning and scheduling are regarded as a key factor for project success. One of the respondents made the following remark about the importance of planning.

> Good planning is a key to effective risk responses. If there is a shortage in any way it is my lack of planning that causes it. The system is a good help if I only have the time to use it as a tool in my planning.

Discussion and conclusions

We found that participation in risk management across the different phases of a project was, as might be expected, largely governed by the actors' roles in those phases. All contractors participated in the production phase and all consultants participated in the design phase. Production was the phase where the majority of respondents participated, while participation in the concept phase was very low. Neither contractors nor clients were sufficiently involved in the concept phase. The design and production phases were identified by all actors as the most important for risk management. In these phases, risk identification, risk assessment and risk response were mostly performed. An important question to investigate further is:

what the actors can gain by participating in all phases of the project? We foresee that participation of all actors in all phases of the construction process leads to more effective risk management through more intensive information and knowledge exchange and earlier identification and assessment of potential project risks.

The results of a survey show that the roles of the actors in the risk management process are strongly associated with their participation in the project's phases. Thus risk identification, risk assessment and risk response were mostly performed in the concept phase by the client; jointly by the client and the consultant in the design phase; and mostly by the contractor in the procurement and production phases. The design and production phases are those where joint risk management was mostly adopted by the actors. We suggest that the procurement phase should play a more important role in joint risk management. Risk management in the project should be based on the actors' shared view of what the risks are and who should carry them. One model might be that the client prepares a view on the risk aspects of the project and the bidding contractor responds with its respective risk analysis. The combined picture of the client's and contractor's risk analyses and shared insights should then form the basis of a conscious risk management process and risk allocation. There is a clear indication that collaboration across all phases of the project increases the probability that a specific risk is managed by the actor who has the best ability to handle it.

Collaboration in risk management was evaluated high by all actors and was most intensive in the production phase. In contrast, the communication of known risks in the procurement phase is low. Collaboration between actors was very strong in risk identification and risk assessment. In risk response, the degree of collaboration decreases significantly according to the views of contractors. This confirms that actors protect their own interests and try to transfer identified risks to others.

Contractors were most active in performing risk identification, risk assessment and risk response systematically on their projects. Moreover, they had the largest influence on risk management in the project overall in the view of other actors. Consultants were regarded as having a very low influence on risk management, perhaps as a result of their lack of familiarity with risk management. It is difficult, however, to generalize as the sample size for consultants is rather small. Clearly though, consultants should be involved more in risk management since design is a significant source of risk. This finding may also be interpreted as a need for consultants to involve other disciplines in order to strengthen or complement their own work in areas such as risk and value.

In terms of small projects, the questions in regard to methods and tools, and how they are used, are short and simple: small projects lack systematic risk management. Even so, contractors, in particular, are involved in continuously handling uncertainty and risk, but rely on an individual's experience and judgment. For small projects, the received view of the risk management process does not fit. Schedules (mostly resource-based), quality assurance and detailed work plans, supplemented by checklists and plain paper, are commonly used. This way of controlling uncertainty and risk cannot be regarded as a form of systematic risk management. Instead, the innate ability of managers is the primary basis upon which project risks are managed. An individual's risk attitude and risk perception are much more important than any formalized system. More research needs to be done to understand how this situation might be changed for the better.

Acknowledgments

The authors gratefully acknowledge the financial support of the Development Fund of the Swedish Construction Industry (SBUF), the Swedish Research Council (Formas) and NCC Construction Sweden.

Notes

1 For ease of comparison and not as a direct monetary equivalent, projects falling in the range of £100,000 to £1 million.
2 Programming is the term used in some countries to refer to this initial phase in the construction process during which the value proposition and need for a project are examined.
3 The scale is between 1 and 4, where 1 = unimportant, 2 = not so important, 3 = fairly important and 4 = very important.
4 Scale is between 1 and 4, where 1 = very bad, 2 = fairly bad, 3 = fairly good and 4 = very good.
5 Scale is between 1 and 4, where 1 = not at all, 2 = little detailed, 3 = fairly detailed, 4 = very detailed.
6 Scale is between 1 and 4, where 1 = very small, 2 = fairly small, 3 = fairly large, 4 = very large.
7 The reference group was made up of representatives from the construction industry having an acceptable level of knowledge about risk management.

23 Quantitative risk management in construction

Kalle Kähkönen

Introduction

Risk management can be seen as a viewpoint for decision-making and other managerial actions. Even so, its precise role, relative importance and practical implications for our work mean that some questions remain unanswered. In general terms, the field of risk management can look confusing and difficult to grasp for a number of reasons. One characteristic of practical risk management is that of highly situation-specific solutions: construction is one such example. 'There are numerous moving parts to an enterprise-wide framework for operational risk management' (Hoffman, 2002) and 'risk management must be tailored to each project – one size does not fit all' (Conrow, 2003). Such statements serve to illustrate the multi-dimensional and challenging nature of real world applications. The classical business process development approach of targeting standardized procedures can produce only limited solutions, which will likely turn out to be a disappointment on live projects.

Risk management needs to combine probabilistic events with current business practices that are usually deterministic in nature. Daily business practice is composed of deterministic objectives, such as budgets, milestones and quality requirements, whereas in the world of risk we face a spectrum of potential outcomes that may or may not happen. Risk management solutions seem to be particularly prone to failure in this regard, because risk analyses are often performed in isolation and, thus, do not have much, if any, impact on decision-making and business performance. The necessary integration of the probabilistic and deterministic worlds has proven to be a challenging task.

A further weakness in the current situation is that risk management is often understood as a separate function to be used in conjunction with other business or project functions; see, for example, PMI (2004) and IPMA (2006). Risk management is a continual process, based on four core tasks: risk identification, analysis, response and control. Continuity of the process is achieved by the cyclic application of these core tasks, but requires a mindset that is different to that used in, say, general management or project planning.

The underlying concepts of risk, opportunity and probability, together with the need for individual judgment, constitute a distinct way of reasoning.

The need for integration can be regarded as a form of pressure intended to present risk management procedures with, and alongside, standard business functions. On the other hand, risk management as a separate function is a force that is pulling it away from other business functions. These two opposing forces, together with solutions that are highly situation-specific, need to be acknowledged as characteristics of risk management.

The aim of this chapter is to present and discuss a set of key elements that seem to form the basis of workable solutions to quantitative risk management.

Literature review

Traditional risk management paradigm

Our thinking, its rationale or lack of it originates from scientific paradigms that have the capacity to present concepts, to structure the part of the world in question and to explain its behavior. A paradigm can be understood as a cluster of beliefs. In terms of research, certain paradigms guide scientists to what should be studied, how research should be undertaken and how the results should be interpreted (Bryman, 1988). It is of importance that researchers try to understand and increase their awareness of the various paradigms that may be affecting their work.

In the field of risk management, it is too easy to anchor one's thinking to the basic model, i.e. a cyclic process of risk identification, analysis, response and control. The application of an action research methodology offers potential for improving the core model, for example by restructuring its content or by introducing new elements to it. Such efforts might be termed explorations according to the traditional paradigm of risk management.

The use of traditional risk management as the starting point is widely documented – see, for example, Boehm (1991), Cooper *et al.* (2005), Chapman and Ward (2002), IPMA (2006), ISO (2003), Nicholas (2004), PMI (2004) and Smith (1999). It would be unfair to label all sources and the work behind them as the promotion of traditional risk management as many go beyond this point. Even so, in most sources, the traditional paradigm is both the basic principle and starting point for explaining the content of risk management.

New emerging risk management paradigm

Quite different views of project risk management are provided by DMO (2006), Pryke and Smyth (2006) and Goldratt (1997). These are examples of how daily project risk management can be understood as a fully embedded dimension within managerial practice and not as a separate function. Grey

(1998) has explained the need to consider risk management in all areas of project management. It is considered that risk management permeating throughout project management is altogether different to the traditional paradigm where the emphasis is on a separate risk management function.

Paradigm shifts tend to be most dramatic in sciences that appear to be stable and mature, as in physics at the beginning of the twentieth century. At that time, physics seemed to be a discipline that was merely providing the last few details of a largely worked-out system of Newtonian mechanics. In 1900, Lord Kelvin famously stated that 'there is nothing new to be discovered in physics now. All that remains is more and more precise measurement'. Five years later, Albert Einstein published his paper on relativity. That paper successfully challenged the scientific principles and rules describing force and motion that had been the dictating theory (and paradigm) for over 300 years.

It would be beneficial if risk management research underwent a paradigm shift from the traditional to a new paradigm. Yet, research and development in risk management have been focused on the traditional paradigm. Competing paradigms can provide ground for fruitful debates and can have profound effects on research design (Bryman and Bell, 2003). The problems and challenges of risk management discussed in the introduction are fundamental and the plain traditional risk management paradigm is too shallow to provide the basis for successful solutions on live projects.

Research into quantitative risk management for construction

Project description and objectives

This chapter is based on a number of case studies where project risk management procedures and related tools have been implemented on live construction projects. The author has developed a software tool for use in this connection. The *Temper System*[1] has been developed for use as a research platform for the exploration of new structures and elements within project risk management. The main objective of the research was to gain improved understanding of the functions that would create a more holistic paradigm compared with traditional risk management.

Research methodology

The approach can be characterized as constructive action research where the main body of work included three implementations of the resultant risk management system (i.e. the *Temper System*) and analysis of the results from its use. The company cases were:

1   Elevator and escalator supplier – an international company carrying out the design and installation of large-scale elevators and escalators in high-rise office and residential buildings.

2   Project management consultant – an international firm with clients in heavy industry where it provides project management services across the whole life cycle of capital investment projects.
3   Risk management organization on a major industrial construction project – the organization faced a challenge when analyze potential problems, their scenarios, back-up plans and information relevant and sufficient to support decision-making.

Quantitative risk analysis was adopted as the common approach and solution for the company studies. Compared with the existing use of qualitative risk estimates, quantitative risk estimates were considered to provide the prospect of links to the commercial and performance aspects of projects, for example cost estimate, budget, schedule and resource utilization plan. Despite this almost self-evident benefit, the use of quantitative risk analysis was found to be very limited on live projects and in the organizations concerned. One likely and significant cause is the imbalance between the amount of information available and the level of detail of estimates. In other words, the accuracy of estimating and reporting needs to match overall understanding of the project and its characteristics.

For each case, the research included the following phases:

1   specification of needs and requirements;
2   adjustments to the *Temper System*;
3   early testing, feedback and additional adjustments; and
4   main test of use on live projects, feedback and documentation.

Research results and industrial impact

The earlier discussion on risk management and paradigms represents the background reasoning forming the main body of the research. The intention was to have more comprehensive coverage over the various challenges of risk management than one originating purely from traditional risk management. Taking different risk management aspects and related challenges as the starting point, a set of elements were then formed as the basis for a practicable risk management solution. These elements were as follows.

1   *Situational applicability*. Flexibility of the solution to cope with various needs throughout the project life cycle.
2   *Conformity with project management culture and processes*. Company practices, its taxonomy and concepts, as well as unspoken rules, need to be incorporated in the solution.
3   *Scope of risk management tasks*. Changes in project conditions and continuity of application of risk management need to be accommodated.

4 *Levels of details in risk modeling.* Essentially, the solution needs to provide several levels of detail for risk modeling. This is the basis for balancing the accuracy of estimates with the amount of information available.
5 *Interfaces for accessing risk models.* These interfaces comprise two components:
   i user interfaces, and
   ii system interfaces.
   User interfaces make risk management procedures effective and easy to apply. System interfaces integrate risk management with other company systems.
6 *Communicativeness.* Characteristics of risk communication need to be understood and incorporated in the solutions.

In general, these elements form a top-down framework for addressing and understanding key aspects of the practical implementation of quantitative risk management. Within the space of this chapter, it is not possible to take a closer look at the fifth element – interfaces for accessing risk models. In the following section, this aspect is addressed by presenting the resultant solution and experiences from its implementation.

Life cycle aspects of risk management in construction

Resultant solution. Building construction has been conventionally understood to comprise several phases from concept to operational use. It is often assumed that the previous phase is fully completed before the next phase starts, or is allowed to start. This can be the case nowadays, for example in the public sector; but in most cases, the different project phases overlap to varying extents and the process as a whole is dynamic, often involving design and construction in parallel, incomplete or changing user requirements, and networked contracting. Additionally, the entire process can include several contractual options that break the whole project into many small supply contracts – a matter of increasing granularity. The nature of these supplies is conditional, wherein the realization of the next supply depends on the previous supply (Figure 23.1).

Utilization on live projects. Traditionally, the main milestones in the project have formed natural steps for carrying out specific risk analyses and related procedures. In addition, we should acknowledge the granular nature of modern construction. Granularity means increasing dynamics, where new parties appear almost continuously such that their contractual obligations and activities can easily produce surprises. This would mean changes in the risk profile of the project: some earlier identified risks may fade away, whereas the severity of others can rise and new risks may appear. The situation is one where the risk profile of the project can be

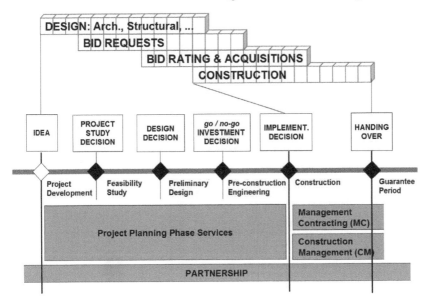

Figure 23.1 Granularity of the construction process with decision points as milestones for specific risk analyses.

changing all the time, so all risk factors possibly affecting it need to be monitored all the time.

Levels of abstraction

Resultant solution. Different levels of abstraction can be an important means for facilitating different tasks that are present in risk management, such as identification of risk titles, additional studies of identified risks and risk status reviews. The structure of these levels provides a framework for this purpose (Figure 23.2).

The level of abstraction is high in terms of qualitative risk knowledge, where risk titles are approached through qualitative descriptions of experiences from past projects. When moving to other levels, the titles and methods become gradually more detailed. The most detailed analyses cover responsive actions, the impact of severe risks and related cost/benefit analyses (Figure 23.3).

Utilization on live projects. In all cases, it was found that the main interest of users was in gaining understanding of the overall picture of risks in the project. Thus, having a complete project risk map turned out to be the level that was most often reached, with users moving toward more detailed levels only when there were compelling reasons, e.g. for additional funding or an increase in budgets (Figure 23.4).

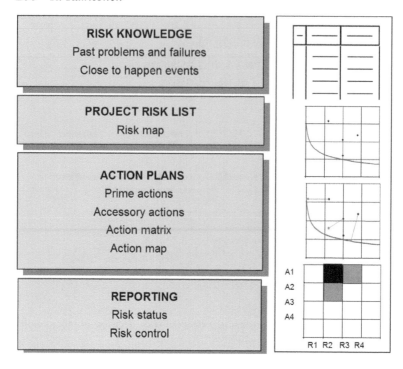

Figure 23.2 Different levels of abstraction from general risk knowledge descriptions to detailed risk reporting.

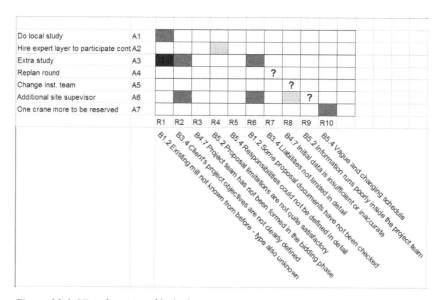

Figure 23.3 Visualization of links between responsive actions and risks.

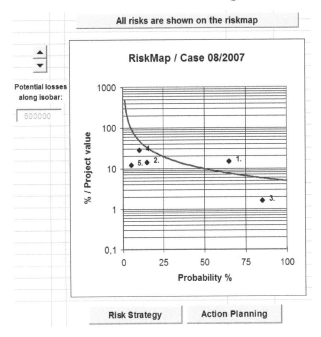

Figure 23.4 The resultant risk map visualizes the details of individual risk estimates.

Quantitative estimates

Resultant solution. The preparation of probability estimates has been a particularly problematic area for quantitative risk management. One of the main results from long-term risk perception studies is that the cognitive structuring of risk using qualitative terminology is similar across nations irrespective of cultural background (Boholm, 1998). This has provided the justification for using natural language expressions when approaching quantitative probability ranges (Hillson, 2005) (Table 23.1).

Table 23.1 Qualitative risk terms with corresponding quantitative ranges

Probability range (%)	Probability value used for calculations (%)	Natural language expression	Numerical score
1 through 14	7	Extremely unlikely	1
15 through 27	21	Low	2
28 through 42	35	Probably not	3
43 through 57	50	50–50	4
58 through 72	65	Probably	5
73 through 86	79	High likelihood	7
87 through 99	93	Almost certainly	9
100	100	Problem already	10

Utilization on live projects. All feedback from the companies was very positive. The proposed model was considered to provide an effective and easy-to-use means for estimating probabilities. Additionally, users found that it was useful to be able to choose more detailed estimates if deemed appropriate (Figure 23.5).

Relevance of risk analysis

Resultant solution. One of the main lessons learned in the early phases from the company cases is the need for impacts arising from risk management to be linked to managerial decision-making and actions. The quantitative estimates are here linked to strategic thinking using a categorization that shows the relative importance of the figures forming the risk map (Figure 23.6). Categorization is based on a risk isobar which is used to classify the data shown on the risk map. Note that the 'risk-taking' level is the same at each point along the risk isobar.

Utilization on live projects. In the company cases, the method for linking individual risk estimates and risk strategy provided a means for discussing the severity of various risks in terms of current market and business climate. It was felt that quantitative risk estimates are not just figures, but have meaning in the context of the company and/or project in question. Completing this successfully provides, perhaps, the most valuable result that can be reached through quantitative risk analysis.

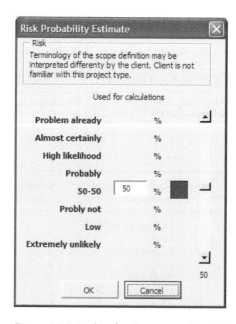

Figure 23.5 Dialog for the input of probability estimates.

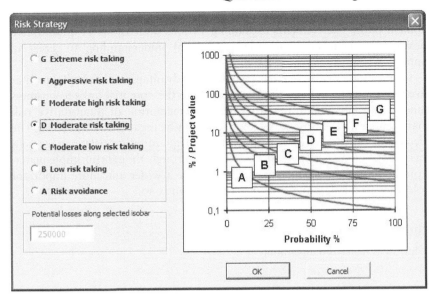

Figure 23.6 Dialog for the selection of risk strategy.

Implementation and exploitation

The *Temper System* software tool is behind the action research and company cases discussed in this chapter. The first version of the tool was originally presented in Kähkönen and Huovila (1995). Since then the tool has been developed through many industrial case studies and implementations. This history and long-term development of the *Temper System* means that it can be considered as a mature tool to be implemented according to the needs of individual companies and projects.

For research purposes, the *Temper System* has provided a platform for exploring potential models to advance the use of risk management within companies. This was also the research approach of the work presented in this chapter. Needs and specifications obtained from the companies have been used as the basis for forming the elemental grounds of proposed solutions, adjustments to the *Temper System* and, finally, testing and validation. The companies referred to earlier in this chapter now use the tool as an integral part of their operations.

Conclusions

Rather than having risk management as a separate function, it needs to be developed into solutions that are increasingly integrated or even embedded within other management functions. This requires an elemental understanding of the content of situation-specific risk management. The work presented in this chapter represents progress along this research and development path. The elements explaining various dimensions of quantitative risk management have provided additional understanding to assist in implementation.

Quantitative risk management has clear advantages over simple qualitative risk estimates. This is due particularly to the potential links to business functions, content and systems. The implementation of quantitative risk management in the three case companies can be regarded as a success. The elemental structuring of the dimensions of quantitative risk management has contributed clearly to this outcome. Through thoughtful implementation, quantitative risk management can have a wider and more important role in project risk management than is presently so.

Note

1 http://cic.vtt.fi/eds/temper.htm.

24 Concentration ratios in the construction market

*Silvio Sancilio, Giuseppe Dibari,
Nicola Costantino and Roberto Pietroforte*

Introduction

Today's construction industry has a global dimension, but also national characteristics that reflect, amongst other things, the concentration of firms in the market, barriers to entry and overall competitiveness. Market concentration ratios can be used as simple, but highly effective, indicators of the competitiveness prevailing in a given market. This chapter addresses the issue of competitiveness in the construction industry from the perspective of market concentration ratios for three example countries. Starting with market structure theory, the chapter introduces the concepts of concentration ratio and competitive markets, as well as the results of previous studies of construction markets. By analyzing the relationship between concentration ratio and market size, three market size categories can be defined: small, medium and large. Sweden, Italy and the USA were selected as the respective examples for these categories. The analysis builds upon homogeneous data in order to make a consistent comparison among the three countries. The issue of subcontracting underscores the point that a construction market may be characterized by an apparent concentration (few main contractors) and, at the same time, fragmentation because of the presence of many subcontractors. In the last part of the chapter, the concentration ratio of US specialist trade contractors is examined to understand the relationship between the concentration ratio and barriers to entry.

Literature review

Market structure models and competitiveness

'The characteristics of a perfectly competitive industry are many small firms with no control over price, producing the same product under conditions of perfect information and no barriers to entry' (de Valence, 2006). Perfect competition means that no producer or consumer has the market power to influence prices. In perfectly competitive markets, participants have no market power. Perfect competition requires five indispensable parameters to enable prices to move instantly to economic equilibrium: atomicity, homogeneity, perfect and complete information, equal access and free entry (Cabral, 2000).

According to Pareto efficiency,[1] perfect competition would lead to a total efficient use of available resources. Agriculture with numerous suppliers, and almost perfectly substitutable products, is an approximation of the perfect competition model. This may have been true in some places at certain times, but in modern times this assertion is hard to prove, because of the complex nature of modern economies – differentiated products, public policies, wealth distribution etc.

A monopoly, where a single firm is the only producer, is the opposite neoclassical model. A firm with such a market power has the individual means to affect either the total quantity or the prevailing price of a product in the market. Different theories have been produced since 1900s to classify existing markets that generally are neither perfect nor monopolistic.

The theory of monopolistic competition pertains to markets with many firms, each having limited control over price, that produce differentiated products and are supported by brand names and marketing with some (often important) barriers to entry. A monopolistically competitive firm acts like a monopolist in that the firm is able to influence the market price of its product by altering its production rate or vice versa (as for monopolists), but it is anyway subject to the competition of similar (alternative) products. In contrast to firms enjoying perfect competition, monopolistically competitive firms produce products that are not perfect substitutes (Chamberlin, 1933).

An oligopoly is created when a few large firms produce identical or differentiated products and operate in a market with significant barriers to entry. Oligopolistic markets are characterized by interactivity: the decisions of one firm influence and are influenced by the decisions of other firms. The strategic planning by oligopolists always involves taking into account the likely responses of other market participants. This puts oligopolistic markets and industries at the highest risk of collusion. The extent of entry barriers is important for defining any market structure model. According to de Valence (2003), these barriers are defined as none, few, significant and very high for perfect competition, monopolistic competition, oligopoly and monopoly respectively. Barriers to entry could play a relevant role in understanding a market structure and its competitiveness. A market with stronger barriers to entry tends to be less competitive, because it contains fewer players and, hence, is more concentrated in structural form (Chiang et al., 2001).

Different definitions of 'barriers to entry' have been proposed over the years. Bain (1956) defined an entry barrier as the set of technology or product conditions that allow incumbent firms to earn economic profits in the long run. Bain stressed three factors which could prevent the entry of a firm: economies of scale, product differentiation and the absolute cost advantages of established firms. Stigler (1968) offered an alternative definition of entry barrier: a cost of producing which must be borne by an entrant, but not by an incumbent.

> In the analysis of entry conditions and barriers to entry, a greater emphasis was initially placed on structural entry conditions, e.g., economies of scale or incumbent cost advantages. Barriers to entry allow us to understand market structure as ... central to market structure is concentration.
>
> (McCloughan, 2004)

Entry barriers to construction

In economics, the market concentration is a function of the number of firms and their respective shares of total production (alternatively total employees) in a market. The ratio is a useful economic tool because it reflects the degree of competition in the market. Alternatively, the ratio expresses the degree of monopoly power exercised by the largest firms. Bain's original concern was based on the intuitive relationship between high concentration and collusion (Tirole, 1988). The concentration ratio is also used as an indicator of the relative size of firms in relation to the industry as a whole. This may also assist in determining the market form of the industry under consideration. The concentration ratio is expressed in the terms Cx, which stands for the percentage of the market sector controlled by the biggest x firms. A disadvantage of this ratio is that it does not indicate the total number of firms that may be operating and competing in an industry. As stated before, market forms could be classified according to their concentration ratio (i.e. perfect competition with very low ratio, monopolistic competition and oligopoly intermediate, and monopoly with very high ratio). In manufacturing, a $C5$ higher than 60 percent usually means a highly concentrated market while a $C5$ lower than 10 percent indicates a very fragmented market. In UK manufacturing, $C5$ is about 20 percent and evidence of a 'correct market' (McCloughan, 2004).

The construction market is certainly not perfect. Due to the large number of firms, construction is generally considered as a highly competitive market and its structure appears to be that of monopolistic competition. It is, however, difficult to classify it in this way. In the case of a design and build competition, for example, proposals on the design and its construction will be different, suggesting the existence of a perfect, competitive market. In some way, the construction market seems to be close to the definition of monopolistic competition. Product homogeneity needs also to be addressed. In the single-product perfect-competition market, firms belonging to the same industry produce a single identical product, which is sold in the same market. The relationship among firms, industry and markets is relatively straightforward. In this context, the industry and market are identical because each has the same group of firms as producers. Nonetheless, this identity does not exist in the construction industry/market, as large firms produce a range of products (many of which are not close substitutes) that are sold in more than one market (de Valence, 2006). In addressing this issue, Gruneberg and Ive (2000) have argued that the construction industry does not produce buildings of different types and its output is not the building itself, but the service of building management. In such a case, services are obviously homogeneous.

The construction industry is divided into many specialist trade sectors that differ in terms of barriers to entry. If companies specializing in tunneling are considered, high barriers exist in terms of required capital for machinery and know-how. In contrast, it is easy to start a painting and decorating business. When protected by entry barriers, firms tend to set their profit margins at higher level and create tacit market collusion in order to keep prices and margins high (Gruneberg and Ive, 2000). These authors have identified six barriers to entry in the construction market.

1   **Economies of scale.** These are the costs associated with the minimum efficient scale of production, below which competition with existing firms would be uneconomical.
2   **Supply chains.** This entry barrier is generally low in construction, but becomes more relevant when it is necessary to manage the activity of different, existing firms.
3   **Incumbents' cost advantages.** This notion is not directly applicable to construction. In manufacturing or distribution, new entrants plan to seize a share of a constant total market from existing suppliers. The entry of a new firm, however, leaves the total demand unchanged. In construction new suppliers plan to poach resources from those firms already in the market.
4   **Private information.** This is the most powerful barrier to entry. Existing producers may take advantage of information not known by entrant firms. Sometimes this knowledge is available, but its use is protected by copyright or patent.
5   **Contestable markets.** Existing firms modify their behavior in order to deter increased competition from potential new entrants. Markets are contestable if there are no sunk costs and entry is perfectly reversible, so that any firm could leave the market recouping the cost of entry.
6   **Client imposed entry barriers to construction contract markets.** This barrier is based on the view of contractor growth as a series of steps of increasing project size and complexity. In this way only few firms can demonstrate past experience on a given project.

Local laws could also represent significant barriers to entry. Firms which plan to enter the public works market in Italy, for example, are required to be SOA (*Società Organismo di Attestazione*) certified. This is a required condition for participating in bidding for public sector contracts. To be SOA certified, a firm has to demonstrate similar prior work experience.

Concentration ratio and market size

The construction market appears to be a highly competitive market with weak entry barriers. Competitiveness depends on market structure and its concentration ratio. This section analyses competitiveness in the construction industry through a multinational comparison based on three different market sizes.

Research methodology

By considering the relationship between the concentration ratio and market size, three market size categories have been defined and investigated: small (up to €50 billion), medium (up to €300 billion) and large (more than €500 billion). Sweden (small), Italy (medium) and the USA (large) have been selected as a representative of these categories. For each country, the market concentration ratio is conventionally measured by using the 5-firm (C5) and the 50-firm (C50) ratios.

Sources of data

The analysis has required two different types of data: national industry turnover and the top five and 50-firm turnover. National industry turnover data have been drawn from the European and national census sources (EUROSTAT, *Statistiska Centralbyrån*, ISTAT and US Census). The top five and 50-firm turnover data have been obtained from national construction bodies or journals (*Sveriges Byggindustrier* for Sweden; Fillea – Cerved Business Information analysis – for Italy; and ENR – Engineering News Record for the USA). A major limitation has been the availability of homogeneous and comparable sources: 2004 data have been used for Sweden and Italy, while 2005 data have been used for the USA. When possible, the C5 and C50 calculations have been performed on the basis of different data sources and also on sector and firm employees.

Multinational concentration ratio comparison

Table 24.1 shows the main data for the three countries.

Table 24.1 Multinational comparison approach showing C5 and C50 values

Swedish Construction Industry	Production value 2004 (€ millions)	Employees 2004
Source: Statistiska Centralbyrån	30,882	187,341
Source: EUROSTAT	30,969	240,502
Source: Statistiska Centralbyrån		
C5 (%)	24.91	16.96
C50 (%)	31.26	n.a.
Source: EUROSTAT		
C5 (%)	24.84	13.21
C50 (%)	31.17	n.a.

Italian Construction Industry		
Source: ISTAT-ANCE	131,893	n.a.
Source: EUROSTAT	186,353	1,833,000
Source: ISTAT-ANCE		
C5 (%)	3.21	n.a.
C50 (%)	9.16	n.a.
Source: EUROSTAT		
C5 (%)	2.27	0.87
C50 (%)	6.48	2.44

USA Construction Industry	Production value 2005 ($ millions)	Employees 2005
Source: US Census	1,143,655	
Source: ENR		
C5 (%)	4.74	n.a
C50 (%)	11.99	n.a

Source: *Costruire, Cerverd,* EUROSTAT, ISTAT, *Statistiska Centralbyrån, Sveriges Byggindustrier,* US Census and ENR.

Results

Swedish market

Sweden has a relatively small construction industry. Even if it is Scandinavian's largest construction market, it still remains small in comparison to that of other EU countries (e.g. France, Germany, the UK, Spain and Italy). The C5 (employment-based) is about 25 percent and higher if compared to that of Italy and USA. The top five Swedish contactors also operate in international markets, but the data collected reflect only national production. Data collected from different sources, show similar values (24.84 percent against 24.91 percent). The C5 employment based figure is lower than the C5 turnover figure. This pattern suggests that the top five firms subcontract part of their work.

Italian market

Several data sources were used for the Italian market. European (EUROSTAT) and national census (ISTAT) data show different values, probably generated by different surveying methods. Sector reports for the top 100 firms have been investigated. Another source was the 2004 Fillea Report (Italian Construction Labor Union). This study provides the gross and national turnover for the top 50 firms, based on the financial audit by Cerved Business Information. Direct employees' data are drawn from Costruire's report for 2004. The value of C5 (turnover based) is 2.27–3.21 percent and it suggests a very highly fragmented market. Figures, especially those related to employment, show low concentration. There are no firms dominating the market. The value of the C50 (turnover based) concentration ratio vary significantly (4.98 percent against 9.16 percent), depending on the two different data sources. The value is lower (2.44 percent) if employment is considered. Similarly to the Swedish case, the data suggest that the top Italian firms are likely to use a large number of subcontractors.

US market

The US Census contains the most complete database for the purpose of this analysis. It provides essential information for government, business, industry and the general public. ENR's report provides turnover information for the top five and 50 firms, but no employment data. US construction is the largest of the markets investigated with the values of C5 and C50 suggesting a perfect market, if compared to the others. Data comparison suggests a low concentration ratio for the construction market. Smaller markets seem to generate higher concentration ratios and, consequently, reveal more market power.

McCloughan's analysis of construction ratios has showed that aggregate concentration is low in the UK construction industry, where the largest 100 private contractors account for 20 percent of activity and 15 percent of employment. In the same study, the market concentration ratio for specialist trade contractors was also examined. Some categories, such as scaffolding (C5 at 56 percent) were characterized by a high concentration ratio. Other studies of the market share by the largest contractors for different countries include Australia (de Valence, 2003), South Korea (Yoon and Kang, 2003), Japan (Woodall, 1996) and Hong Kong (Chiang et al., 2001).

The employment-based concentration ratio is lower than that based on turnover (e.g. for Italy the employment based C5 is 0.87 percent against 2.27 percent). This pattern suggests that very large contractors subcontract a significant proportion of their work. The concentration ratio appears to decrease as the market size increases. Understanding of the relationship between market size and concentration ratio requires further investigation according to the concentration ratios within the various sub-markets of construction.

Concentration ratio in US construction

The lack of homogeneous construction outputs means that the relationship between the concentration ratio and nature of a specific sector has to be examined in order to show the effect of entry barriers in a given market sector. The results are shown in Table 24.2. Four different US construction market sectors were considered as representative of market outputs according to required know-how, initial capital investment, supply chain and labor cost. The sectors considered were electrical services, glazing (curtain walls), painting and transportation (highways and bridges). The values of the C5, C20 and C50 indexes have been calculated where data were available.

According to the initial assumptions, high entry barriers (in terms of capital investment) lead to higher C5 indexes, as in the case of glazing and transportation. A more concentrated market entails relatively higher expenses for acquiring the goods necessary for production. This pattern suggests that if the final output of a given sector results more from off-site than site production operations, its concentration ratio increases for the required use of machinery.

As Figure 24.1 shows, concentration ratios follow a linear distribution and increase proportionally as the number of firms grows.

Table 24.2 C5, C20, C50 in US construction sectors

Construction sector	Electrical	Glazing	Painting	Transportation
Production value 2002 ($ millions)	$83,377	$6,398	$16,958	$83,355
C5	5.65%	9.97%	2.32%	12.69%
C20	9.38%	15.35%	3.69%	20.74%
C50	13.36%	n.a.	n.a.	29.37%
Cost of materials, components, supplies and fuels against value of business	32.11%	41.09%	22.97%	33.67%
Cost of employees against value of business	35.46%	27.48%	35.47%	18.78%

Source: US Census, ENR.

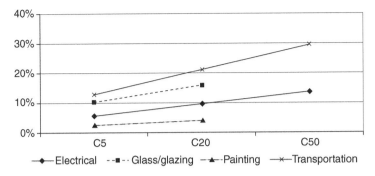

Figure 24.1 C5, C20, C50 in the US construction sectors.

Conclusions

This chapter has addressed the economic characteristics of construction markets with the emphasis on barriers to entry. In this last regard, several theories have been outlined. The market concentration ratio of three differently sized construction markets – Sweden, Italy and the USA – have been calculated. The value of the ratios seems to decrease as market size increases. The study, in addition, addressed some of the sectors that make up the US construction industry. Data show that specialist sectors, such as the transportation and glazing (curtain walling) markets, have relatively higher concentration ratios. It is argued that these two sectors have higher entry barriers because of the requirements in terms of capital investment and know-how. Future studies based on a larger sample of national data should verify these preliminary findings, namely the inverse relationship between market size and concentration ratio, as well as the relationship between type of market sub-sector and concentration ratio.

Note

1 Sometimes referred to as Pareto optimality.

References

Aaltonen, K. and Sivonen, R. (2009) Response strategies to stakeholder pressures in global projects. *International Journal of Project Management*, **27**(2), 131–41.

Aaltonen, P. and Ikavalko, H. (2002) Implementing strategies successfully. *Integrated Manufacturing Systems*, **13**(6), 415–18.

AbouRizk, S. and Shi, J.S. (1994) Automated construction-simulation optimization. *J. Construction Engineering and Management*, **120**(2), 374–85.

AbouRizk, S., Halpin, D. and Lutz, J.D. (1992) State of the art in construction simulation. *Proc. 1992 Winter Simulation Conference*. 13–16 December, Arlington, VA, 1271–7.

Abudayyeh, O., Dibert-DeYoung, A. and Jaselskis, E. (2004) Analysis of trends in construction research: 1985–2002. *J. Construction Engineering and Management*, **130**(3), 433–9.

Ackerman, M.S. (2000) The intellectual challenge of CSCW: the gap between social requirements and technical feasibility. *Human-Computer Interaction*, **15**(2–3), 179–203.

Ackoff, R.L. (1979) The future of operational research is past. *J. Operational Research Society*, **30**, 93–104.

Ahlenius, E. (1999) Om lönsam och effektiv riskhantering [About profitable and effective risk management]. *Väg- och Vattenbyggaren*, (1:99), 26–9 (in Swedish).

Ahmed, S., Ahmed, R. and Saram, D. (1999) Risk management trends in the Hong Kong construction industry: a comparison of contractors' and owners' perceptions. *Engineering, Construction and Architectural Management*, **6**(3), 225–34.

Akintoye, A.S. and MacLeod, M.J. (1997) Risk analysis and management in construction. *International Journal of Project Management*, **15**(1), 31–8.

Allio, M.K. (2005) A short, practical guide to implementing strategy. *The Journal of Business Strategy*, **26**(4), 12–21.

Anvuur, A.M. (2008) *Cooperation in Construction Projects: Concept, Antecedents and Strategies*. Doctoral thesis, Department of Civil Engineering, The University of Hong Kong, Hong Kong.

APM (2004) *Project Risk Analysis and Management Guide*. Association for Project Management, APM Publishing, High Wycombe.

Aravena, J.L. (2005) Speech act theories for computer aided collaborative design. *International Journal of Design Computing*. Online journal, available at: www.arch.usyd.edu.au/kcdc/journal/vol3/dcnet/avarena/abstract.html (last accessed December 2008).

Arrighetti, A., Bachmann, R. and Deakin, S. (1997) Contract law, social norms and inter-firm cooperation. *Cambridge Journal of Economics*, **21**(2), 171–95.

Arthur, W.B. (1994) Inductive reason and bounded rationality (The El Farol Problem). *American Economic Review*, **84**(2), 406–11.

Arthur, W.B. (2000) Cognition: the black box of economics. In D. Colander (ed.) *The Complexity Vision and the Teaching of Economics*. Edward Elgar, Northampton, MA.

Arthur, W.B. (2007) The structure of invention. *Research Policy*, **36**(2), 274–87.

Asphalt Institute (2007) *MS-4 The Asphalt Handbook*. The Asphalt Institute, Lexington, KY.

Axelsson, R. (1981) *Rationell organisation* [Rational Organisations]. Norstedts juridik, Stockholm (in Swedish).

Baccarini, D. and Archer, R. (2001) The risk ranking of projects: a methodology. *International Journal of Project Management*, **19**(3), 139–45.

Bachmann, R., Knights, D. and Sydow, J. (2001) Editorial: trust and control in organization relations. *Organization Studies*, **22**(3), 1–4.

Bailey, A. and Johnson, G. (1992) How strategies develop in organizations. In D. Faulkner and G. Johnson (eds) *The Challenge of Strategic Management*. Kogan Page, London.

Bain, J. (1956) *Barriers to New Competition*. Harvard University Press, Cambridge, MA.

Bakis, N., Kagiouglou, M., Aouad, G., Amaratunga, D., Kishk, M. and Al-Hajj, A. (2003) An integrated environment for life cycle costing in construction. In R. Amor (ed.) *Construction IT Bridging the Distance. Proc. CIB W78 20th International Conference on Information Technology in Construction*, New Zealand, 23–25 April. CIB Report: Publication 284, CIB, Rotterdam, the Netherlands.

Ball, M., Farshchi, M. and Grilli, M. (2000) Competition and the persistence of profits in the UK construction industry. *Construction Management and Economics*, **18**(7), 733–45.

Ballard, G. (1994) *The Last Planner*. Northern California Construction Institute, Monterey, CA.

Ballard, G. (2005) Construction: one type of project-based production system. In *Proc. SCRI Forum Event Lean Construction: The Next Generation*, 19 January. SCRI, University of Salford, 14.

Baloi, D. and Price, A.D.F. (2003) Modelling global risk factors affecting construction cost performance. *International Journal of Project Management*, **21**(4), 261–9.

Barber, R.B. (2005) Understanding internally generated risks in projects. *International Journal of Project Management*, **23**(8), 584–90.

Barker, K. (2003) *Review of Housing Supply, Securing our Future Housing Needs Interim Report – Analysis*. HM Treasury, London.

Barker, K. (2004) *Review of Housing Supply, Delivering Stability: Securing our Future Housing Needs Final Report – Recommendations*. HM Treasury, London.

Barlow, J. (1993) Controlling the housing land market: some examples from Europe. *Urban Studies*, **30**(7), 1129–49.

Barlow, J. and King, A. (1992) The state, the market, and competitive strategy: the housebuilding industry in the United Kingdom, France and Sweden. *Environment and Planning A*, **24**(3), 381–400.

Barrett, P.S. (1999) Good practice in briefing: the limits of rationality. *Automation in Construction*, **8**(6), 633–42.

Barrett, P.S. and Sexton, M.G. (2006) Innovation in small, project-based construction firms. *British Journal of Management*, **17**(4), 331–46.

Baumann, H., Brunklaus, B., Gluch, P., Kadefors, A., Stenberg, A.-C. and Thuvander, L. (2003) *Byggsektorns miljöbarometer 2002* [The Environmental Barometer for the Construction Sector 2002]. CMB-report, ESA Report 2003:2. Chalmers University of Technology, Gothenburg.

Bekkering, T., Glas, H., Klaassen, D. and Walter, J. (2004) *Management van processen* [Management Processes]. Het Spectrum, Utrecht (in Dutch).

Bertelsen, N.H. (2004) *Bedre produktivitet ved renovering. Forsøg med planlægning, styring og opfølgning på sager* [Better Productivity in Renewal – Case-studies on Planning, Managing and Follow-up]. By og Byg Dokumentation 061. Danish Building Research Institute, Copenhagen (in Danish).

Bertelsen, N.H. (2005) *Den selvstyrende byggeplads* [The Self-governing Construction Site]. SBI Report 2005:11. Danish Building Research Institute, Copenhagen (in Danish).

Bertelsen, N.H. (2007) *How to Reduce Building Costs – Danish Initiatives.* Project 741–079, memo 26/1 2007. Danish Building Research Institute, Copenhagen.

Bijker, W.E. (ed.) (1987) *The Social Construction of Technological Systems: New Directions in the Sociology and History of Technology.* MIT Press, Cambridge, MA.

Bijker, W.E. and Law, J. (1992) *Shaping Technology/Building Society.* MIT Press, Cambridge, MA.

Blismas, N., Sher, W., Thorpe, A. and Baldwin, A. (2004) A typology for clients' multi-project environments. *Construction Management and Economics*, **22**(4), 357–71.

Blockley, D. and Godfrey, P. (2000) *Doing it Differently – Systems for Rethinking Construction.* Thomas Telford, London.

Boehm, B.W. (1991) Software risk management: principles and practices. *IEEE Software*, **8**(1), 32–41.

Boholm, Å. (1998) Comparative studies of risk perception: a review of twenty years of research. *J. Risk Research*, **1**(2), 135–63.

Borgbrant, J. (1990) Strategisk dialog 2 [Strategic dialogue 2]. *Natur och kultur*, Stockholm (in Swedish).

Bossink, B. (2004) Managing drivers of innovation in construction networks. *J. Construction Engineering and Management*, **130**(3), 337–45.

Bouvet, D., Froumentin, M. and Garcia, G. (2001) A real-time localization system for compactors. *Automation in Construction*, **10**(4), 417–28.

Boverket (2002) *Bostadsbyggandet i planeringen* [Production of Housing and the Planning Process]. Boverket [National Board of Housing, Building and Planning], Karlskrona (in Swedish).

Bowden, S., Dorr, A., Thorpe, T. and Anumba, C. (2006) Mobile ICT support for construction process improvement. *Automation in Construction*, **15**(5), 664–76.

Bowen, P., Pearl, R. and Akintoye, A. (2007) Professional ethics in the South African construction industry. *Building Research & Information*, **35**(2), 189–205.

Bradach, J.L. and Eccles, R.G. (1989) Price, authority and trust: from ideal types to plural forms. In W.R. Scott (ed.) *Annual Review of Sociology*, **15**, 97–118.

Brahy, S. (2006) Six solutions pillars for successful cultural integration of international M&As. *J. Organisational Excellence*, **25**(4), 53–63.

Braster, J.F.A. (2000) *De kern van casestudy's* [The Essence of Case Studies]. Van Gorcum, Assen.

Bresnen, M. and Marshall, N. (2000) Building partnerships: case studies of client–contractor collaboration in the UK construction industry. *Construction Management and Economics*, **18**(7), 819–32.

Bresnen, M., Goussevskaia, A. and Swan, J. (2004) Embedding new management knowledge in project-based organizations. *Organization Studies*, **25**(9), 1535–55.

Bresnen, M., Goussevskaia, A. and Swan, J. (2005) Editorial: managing projects as complex social settings. *Building Research & Information*, **33**(6), 487–93.

Bridge, A. and Tisdell, C. (2004) The determinants of the vertical boundaries of the construction firm. *Construction Management and Economics*, **22**(8), 807–25.

Briscoe, G.H., Dainty, A.R.J, Millett, S.J. and Neale, R.H. (2004) Client-led strategies for construction supply chain improvement. *Construction Management and Economics*, **22**(2), 193–201.

Brooks, I. (2006) *Organizational Behaviour: Individuals, Groups and Organization*. 3rd edn. Financial Times/Prentice Hall, London.

Bruil, A.W. (2004) Sturing en procesmanagement [Steering and process management]. In A.W. Bruil, F.A.M. Hobma, G.J. Peek and G. Wigmans (eds) *Integrale gebiedsontwikkeling: Het stationsgebied 's-Hertogenbosch* [Integrated Area Development: the Station Area of 's-Hertogenbosch]. Amsterdam, 259–79 (in Dutch).

Bryman, A. (1988) *Quantity and Quality in Social Research*. Routledge, London.

Bryman, A. (2008) *Social Research Methods*. 3rd edn. Oxford University Press, Oxford.

Bryman, A. and Bell, E. (2003) *Business Research Methods*. Oxford University Press, Oxford.

Buchanan, J.T., Henig, E.J. and Henig, M.I. (1998) Objectivity and subjectivity in the decision-making process. *Annals of Operations Research*, **80**(0), 333–45.

Buono, A.F. and Bowditch, J.L. (1989) *The Human Side of Mergers and Acquisitions*. Jossey-Bass, San Francisco, CA.

Cabral, L.M.B. (2000) *Introduction to Industrial Organisation*. MIT Press, Boston, MA.

Carlsen, A., Klev, R. and von Krogh, G. (eds) (2004) *Living Knowledge. The Dynamics of Professional Service Work*. Palgrave Macmillan, Basingstoke.

Carn, N., Rabianski, J., Racster, R. and Seldin, M. (1998) *Real Estate Market Analysis: Techniques and Applications*. Prentice-Hall, Englewood Cliffs, NJ.

Cassel, P. and Hjelmfeldt, M. (2001) *Marknad och ledning* [Market and Management]. Svensk byggtjänst [Swedish Building Centre], Stockholm (in Swedish).

Chadboum, B.A., Newcomb, D., Voller, V.R., DeSombre, R.A., Luoma, J.A. and Timm, D. (1998) *An Asphalt Paving Tool for Adverse Conditions*. Final Report MN/DOT 1998-18. Department of Civil Engineering, University of Minnesota.

Chaffee, E.E. (1985) Three models of strategy. *Academy of Management Review*, **10**(1), 89–98.

Chamberlin, E. (1933) *The Theory of Monopolistic Competition*. Harvard University Press, Boston, MA.

Chapman, C.B. and Ward, S.C. (2002) *Managing Project Risk and Uncertainty: a Constructively Simple Approach to Decision Making*. Wiley, Chichester.

Chapman, C.B. and Ward, S.C. (2003) *Project Risk Management: Processes, Techniques and Insights*. 2nd edn. Wiley, Chichester.

Checkland, P. (1981) *Systems Thinking, Systems Practice*. Wiley, New York, NY.

Cherns, A.B. and Bryant, D.T. (1984) Studying the client's role in construction management. *Construction Management and Economics*, 2(2), 177–84.

Cheung, F.Y.K., Rowlinson, S., Jefferies, M. and Lau, E. (2005) Relationship contracting in Australia. *J. Construction Procurement: Special Issue on Trust in Construction*, 11(2), 123–35.

Chiang, Y.H., Tang, B.S. and Leung, W.Y. (2001) Market structure of the construction industry in Hong Kong. *Construction Management and Economics*, 19(7), 675–87.

Child, J. (1972) Organizational structure, environment and performance: the role of strategic choice. *Sociology*, 6(1), 1–22.

Chisholm, R.F. and Elden, M. (1993) Features of emerging action research. *Human Relations*, 46(10), 275–98.

Choi, J. and Minchin, R.E. (2006) Workflow management and productivity control for asphalt pavement operations. *Canadian Journal of Civil Engineering*, 33(8), 1039–49.

Cicmil, S. (2006) Understanding project management practice through interpretative and critical research perspectives. *Project Management Journal*, 37(2), 27–37.

Cicmil, S. and Hodgson, D. (2006) New possibilities for project management theory: a critical engagement. *Project Management Journal*, 37(3), 111–22.

Cicmil, S., Williams, T., Thomas, J. and Hodgson, D. (2006) Rethinking project management: researching the actuality of projects. *International Journal of Project Management*, 24(8), 675–86.

Clarke, A.E. and Fujimura, J. (eds) (1992) *The Right Tools for the Job: at Work in Twentieth-century Life Sciences*. Princeton University Press, Princeton, NJ.

Clegg, S.R. (2005) *Talking Construction into Being*. Inaugural address, Vrije Universiteit, Amsterdam.

Clegg, S.R., Pitsis, T.S., Rura-Polley, T. and Marosszeky, M. (2002) Governmentality matters: designing an alliance culture of inter-organizational collaboration for managing projects. *Organization Studies*, 23(3), 317–37.

Cleland, D.I. and Ireland, L.R. (2007) *Project Management: Strategic Design and Implementation*. 5th edn. McGraw-Hill, New York, NY.

Coase, R. (1937) The nature of the firm. *Economica*, 4(16), 386–405.

Cohen, M.D., March, J.G. and Olsen, J.P. (1972) A garbage can model of organizational choice. *Administrative Science Quarterly*, 17, 1–25.

Cole, R.J. and Sterner, E. (2000) Reconciling theory and practice of life-cycle costing. *Building Research & Information*, 28(5/6), 368–75.

Collingridge, D. (1980) *The Social Control of Technology*. St. Martin's Press, New York, NY.

Collins, J.C. and Porras, J.I. (1994) *Built to Last: Successful Habits of Visionary Companies*. Harper Business, New York, NY.

Conrow, E.H. (2003) *Effective Risk Management: Some Keys to Success*. American Institute of Aeronautics and Astronautics, Reston, VA.

Cooper, D., Grey, S., Raymond, G. and Walker, P. (2005) *Managing Risk in Larger Projects and Complex Procurements*. Wiley, Chichester.

Corboy, M. and O'Corrbui, D. (1999) The seven deadly sins of strategy. *Management Accounting*, 77(10), 29–30.

Corner, J., Buchanan, J. and Henig, M. (2001) Dynamic decision problem structuring. *J. Multi-Criteria Decision Analysis*, 10(3), 112–41.

Cox, A. and Ireland, P. (2002) Managing construction supply chains: the common sense approach. *Engineering, Construction and Architectural Management*, 9(5/6), 409–18.

Cox, A. and Townsend, M. (1998) *Strategic Procurement in Construction: Towards Better Practice in the Management of Construction Supply Chains*. Thomas Telford, London.

Cyert, R.M. and March, J.G. (1963) *A Behavioral Theory of the Firm*. Prentice-Hall, Englewood Cliffs, NJ.

Czarniawska, B. and Joerges, B. (1996) Travels of ideas. In B. Czarniawska and G. Sevón (eds) *Translating Organizational Change*. Walter de Gruyter, Berlin, 13–48.

Dahlbom, B. and Mathiassen, L. (1993) *Computers in Context: the Philosophy and Practice of Systems Design*. Blackwell, Oxford.

Dainty, A., Moore, D. and Murray, M. (2006) *Communication in Construction: Theory and Practice*. Taylor & Francis, Oxford.

Dancy, J. (1985) *An Introduction to Contemporary Epistemology*. Blackwell, Oxford.

Das, T.K. and Teng, B.S. (2001) Trust, control and risk in strategic alliances: an integrated framework. *Organization Studies*, 22(2), 251–83.

Davenport, T.H. and Prusak, L. (1998) *Working Knowledge – how Organizations Manage what they Know*. Harvard Business School Press, Boston, MA.

Davis, F. (1989) Perceived usefulness, perceived ease of use, and user acceptance of information technology. *MIS Quarterly*, 13(3), 319–40.

De Geus, A.P. (1988) Planning as learning. *Harvard Business Review*, 66(2), 70–4.

De Leeuw, A.C. (2002) *Bedrijfskundig management – primair process, strategie en organisatie* [Business Management – the Primary Process, Strategy and Organisation]. Koninklijke Van Gorcum, Assen (in Dutch).

De Ridder, H.A.J., Klauw, R.A. van der and Vrijhoef, R. (2002) *Het nieuwe bouwen in Nederland* [A New Way of Constructing in the Netherlands]. TNO, Delft (in Dutch).

De Valence, G. (2003) Market structure, barriers to entry and competition in construction markets. In *Proc. Joint CIB W55/W65/W107 International Symposium Knowledge Construction*, Singapore, 819–28.

De Valence, G. (2006) Competition and the characteristics of construction markets. In *Proc. Construction in the XXI Century: Local and Global Challenges, Joint 2006 CIB W065/W055/W086 Working Commissions International Symposium*, 18–20 October, Rome, 82–3.

Deetz, S. (2003) Reclaiming the legacy of the linguistic turn. *Organization*, 10(3), 421–9.

Del Caño, A. and De La Cruz, M. (2002) Integrated methodology for project risk management. *J. Construction Engineering and Management*, 128(6), 473–85.

Deming, W.E. (1986) *Out of the Crisis*. Cambridge University Press, Cambridge.

Descartes, R. (1637) *Discourse on the Method* [Discours de la méthode]. Translated from French into Danish by Brøndal, V. and Valdemar, H. (1996) Gyldendal, Copenhagen.

Desouza, K.C. and Awazu, Y. (2006) Knowledge management at SMEs: five peculiarities. *J. Knowledge Management*, **10**(1), 32–43.

Dey, I. (1993) *Qualitative Data Analysis: a User-friendly Guide for Social Scientists.* Routledge, London.

DMO (2006) *Competency Standard for Complex Project Managers – Version 2.0.* Defence Materiel Organisation, Department of Defence, Canberra, ACT.

Doherty, T.A. (1988) Resolving culture conflicts during the merger. *Chief Executive*, **45**, 18–22.

Donaldson, T. and Preston, L. (1995) The stakeholder theory of the corporation: concepts, evidence, and implications. *Academy of Management Review*, **20**(1), 65–91.

Dorée, A.G. (2004) Collusion in the Dutch construction industry: an industrial organization perspective. *Building Research & Information*, **32**(2), 146–56.

Dorée, A.G. and de Ridder, H.A.J. (2003) *Oriëntaties op de toekomst* [Orientations on the Future]. AVVB, Moordrecht (in Dutch). Online, available at: http://doc.utwente.nl/58060/1/bouwen_pos_dyn-1.pdf.

Dorée, A.G. and ter Huerne, H.L. (2005) *Development of Tools and Models for the Improvement of Asphalt Paving Process.* Workshop proceedings. Department of Civil Engineering and Management, University of Twente, Enschede.

Dreschler, M. and de Ridder, H.A.J. (2006) Results of an explorative research into value quantification methods. In D. Amaratunga *et al.* (eds) *Proc. 6th International Postgraduate Research Conference.* 6–8 April, University of Salford and Delft University of Technology, Delft, the Netherlands, 588–96.

Drucker, P.F. (1963) *The Practice of Management.* Heinemann, London.

Drucker, P.F. (2003) *The New Realities: in Government and Politics.* Random House, New York, NY.

Dubois, A. and Gadde, L.-E. (2002) The construction industry as a loosely coupled system: implications for productivity and innovation. *Construction Management and Economics*, **20**(7), 621–31.

Dulaimi, M., Ling, F. and Bajracharya, A. (2002) Enhancing integration and innovation in construction. *Building Research & Information*, **30**(4), 237–47.

Eccles, R.G. (1981a) Bureaucratic versus craft administration: the relationship of market structure to the construction firm. *Administrative Science Quarterly*, **26**(3), 449–69.

Eccles, R.G. (1981b) The quasi-firm in the construction industry. *J. Economic Behavior and Organization*, **2**(4), 335–57.

Edum-Fotwe, F.T. and Azinim, M. (2006) Risk management practices of construction project staff: preliminary lessons. *Proc. Construction in the XXI Century: Local and Global Challenges, Joint 2006 CIB W065/W055/W086 Working Commissions International Symposium*, 18–20 October, Rome, 20–1.

Edvinsson, L. and Malone, S.M. (1997) *Intellectual Capital: Realizing Your Company's True Value by Finding Its Hidden Brainpower.* HarperCollins, New York, NY.

Egan, J. (1998) *Rethinking Construction.* Department of the Environment, Transport and the Regions, Stationery Office, London.

Egeberg, M. (1984) *Organisasjonsutformning i effentlig virksamhet* [Structure in Public Organisations]. Aschehoug/Tanum-Norli, Oslo (in Norwegian).

Eisenhardt, K.M. (1989) Building theories from case study research. *Academy of Management Review*, **14**(4), 532–50.

Eisenhardt, K.M. and Bourgeois III, L.J. (1988) Politics of strategic decision making in high-velocity environments: toward a midrange theory. *Academy of Management Journal*, 31(4), 737–70.

Eisenhardt, K.M. and Zbaracki, M.J. (1992) Strategic decision making. *Strategic Management Journal*, Special Issue: Fundamental Themes in Strategy Process Research, 13, 17–37.

Ekman, B. (1970) *Beslutsrationalisering* [Rationalisation of Decisions]. Almqvist & Wiksell, Uppsala (in Swedish).

Emblemsvåg, J. (2003) *Life-cycle Costing: Using Activity-based Costing and Monte Carlo Methods to Manage Future Costs and Risks*. Wiley, New York, NY.

Engwall, Å. (2001) *Dynamics in Refurbishment: a Study of Production Processes and Human Interactions in a Commercial Environment*. Licentiate dissertation, Department of Civil Engineering, Luleå University of Technology, Luleå.

ENR (2005) 2005 top contractors. *Engineering News Record*, 16 May, 4–17.

Ericsson, B. and Johansson, B.-M. (1994) *Bostadsbyggandet i idé och praktik – om kunskap och föreställningar inom byggsektorn* [Housebuilding in Concept and Practice – Knowledge and Performance in the Building Sector]. Department of Sociology, Lund University, Lund (in Swedish).

Ericsson, L.E., Liljelund, L.E., Sjöstrand, M., Uusmann, I., Modig, S., Ärlebrant, Å. and Högrell, O. (2002) *Skärpning gubbar: om konkurrensen, kvaliteten, kostnaderna och kompetensen i byggsektorn* [About Competition, Quality, Cost and Competence in the Construction Sector]. Byggkommissionens betänkande [Reflections of the Construction Commission], SOU 2002:115, Fritzes, Stockholm.

Fairhurst, G.T., Jordan, J.M. and Neuwirth, K. (1997) Why are we here? Managing the meaning of an organizational mission statement. *J. Applied Communication Research*, 25(4), 243–63.

Faulkner, D. and Campbell, A. (eds) (2003) *The Oxford Handbook of Strategy. A Strategy Overview and Competitive Strategy*. 1, Oxford University Press, Oxford.

Fawcett, S.E. and Magnan, G.M. (2002) The rhetoric and reality of supply chain integration. *International Journal of Physical Distribution & Logistics Management*, 32(5), 339–61.

Femenías, P. (2004) *Demonstration Projects for Sustainable Building: Towards a Strategy for Sustainable Development in the Building Sector based on Swedish and Dutch experience*. Doctoral thesis, Chalmers University of Technology, Gothenburg.

Fincham, R. (2002) Narratives of success and failure in systems development. *British Journal of Management*, 13(1), 1–14.

Flanagan, R. and Jewell, C.A. (2005) *Whole Life Appraisal for Construction*. Blackwell, Oxford.

Flanagan, R. and Norman, G. (1993) *Risk Management and Construction*. Blackwell, Oxford.

Flanagan, R., Norman, G., Meadows, J. and Robinson, G. (1989) *Life Cycle Costing – Theory and Practice*. Blackwell, Oxford.

Fleck, J. (1994) Learning by trying: the implementation of configurational technology. *Research Policy*, 23(6), 637–52.

Flyvbjerg, B. (2001) *Making Social Science Matter: Why Social Inquiry Fails and How It Can Succeed Again*. Cambridge University Press, Cambridge.

Foos, T., Schum, G. and Rothenberg, S. (2006) Tacit knowledge transfer and the knowledge disconnect. *J. Knowledge Management*, **10**(1), 6–18.

Foster, C. (1964) Competition and organization in building. *J. Industrial Economics*, **12**(3), 163–74.

Foster, S.T. (2004) *Managing Quality – an Integrative Approach*. 2nd edn. Prentice-Hall, Upper Saddle River, NJ.

Frambach, R.T. (1993) An integrated model of organizational adoption and diffusion of innovations. *European Journal of Marketing*, **27**(5), 22–41.

Fredrickson, J.W. (1983) Strategic process research: questions and recommendations. *Academy of Management Review*, **8**(4), 565–75.

Fredrickson, J.W. (1986) The strategic decision process and organizational structure. *Academy of Management Review*, **11**(2), 280–97.

Freeman, R.E. (1984) *Strategic Management – A Stakeholder Approach*. Pitman, Publishing, Marshfield, MA.

Fristedt, S. and Ryd, N. (2003) *Ju förr desto bättre: programmarbete i tidigt skede av byggprocessen* [The Sooner the Better: Briefing in the Early Stage of the Building Process]. Svenska kommunförbundet [Swedish Association of Local Authorities and Regions], Stockholm (in Swedish).

Fujimura, J. (1992) Crafting science: standardized packages, boundary objects, and 'translation'. In A. Pickering (ed.) *Science as Practise and Culture*. University of Chicago Press, Chicago, IL, 168–211.

Galbraith, J. (1973) *Designing Complex Organizations*. Addison-Wesley, Reading, MA.

Gann, D., Wang, Y. and Hawkins, R. (1998) Do regulations encourage innovation? The case of energy efficiency in housing. *Building Research & Information*, **26**(4), 280–96.

Gareth, R.J. (1995) *Organizational Theory: Text and Cases*. Addison-Wesley Publishing, Reading, MA.

Gehner, E. (2008) *Knowingly Taking Risk: Investment Decision Making in Real Estate Development*. Doctoral thesis, Eburon Academic Publishers, Delft.

Gehner, E. and de Jonge, H. (2005) A cognitive perspective on risk management in real estate development. In Y. Wang *et al.* (eds) *International Conference on Construction and Real Estate Management*. China Architecture & Building Press, Penang, Malaysia, 739–44.

Gehner, E., Halman, J.I.M. and de Jonge, H. (2006) Risk management in the Dutch real estate development sector: a survey. In D. Amaratunga *et al.* (eds) *Proc. 6th International Postgraduate Research Conference*. 6–8 April, University of Salford and Delft University of Technology, Delft, the Netherlands, 541–52.

Gerdemark, T. (2000) *Kundorientering – oklarheter på den självklara vägen* [Customer-orientation – Ambiguities in an Obvious Way]. Department of the Built Environment, University of Gävle, Gävle (in Swedish).

Geurts, J.L.A. and Vennix, J.A.M. (eds) (1989) Verkenningen in beleidsanalyse: theorie en praktijk van modelbouw en simulatie [Explorations of policy analysis: theory and practice of modelling and simulation]. In F. van Hamelen (ed.) *European Conference on Artificial Intelligence*. Zeist, Kerckebosch, IOS Press, Amsterdam (in Dutch).

Geurts, J.L.A. and Weggeman, M.P. (1992) Strategische beslissingsanalyse voor R&D-projecten [Strategic decision analysis for R&D projects]. *Bedrijfskunde*, **64**(1), 68–80 (in Dutch).

Ghemawat, P. (1999) *Strategy and the Business Landscape*. Addison-Wesley, Reading, MA.

Glaser, B.G. and Strauss, A.L. (1967) *The Discovery of Grounded Theory: Strategies for Qualitative Research*. Aldine, New York, NY.

Gluch, P. (2005) *Building Green – Perspectives on Environmental Management in Construction*. Doctoral thesis, Department of Civil and Environmental Engineering, Chalmers University of Technology, Gothenburg.

Gluch, P., Brunklaus, B., Johansson, K., Lundberg, Ö., Stenberg, A.-C. and Thuvander, L. (2006) *Miljöbarometern för bygg- och fastighetssektorn 2006 – en kartläggning av sektorns miljöarbete* [The Environmental Barometer for the Construction Sector 2006 – a Survey]. CMB-report. Chalmers University of Technology, Gothenburg.

Goldratt, E.M. (1997) *Critical Chain*. The North River Press, Great Barrington, MA.

Gonzalez-Diaz, M., Arrunada, B. and Fernandez, A. (2000) Causes of subcontracting: evidence from panel data on construction firms. *Economic Behavior & Organization*, 42(2), 167–87.

Goovaerts, A., de Ridder, H.A.J. and Hombergen, L.P.I.M. (2004) *Afwegingsmethodieken – Een overzicht van bestaande methodieken* [Evaluation Techniques – an Overview of Existing Methodologies]. KCBPI, Delft (in Dutch).

Gottlieb, S.G. and Bertelsen, N.H. (2006) *Byggestyring for fagentreprenører. Erfaringer og ideudvikling med baggrund i murerfaget* [Construction Control for Subcontractors]. SBI Report 2006:10. Danish Building Research Institute, Copenhagen (in Danish).

Granovetter, M.S. (1985) Economic action and social structure: a theory of embeddedness. *American Journal of Sociology*, 91(3), 481–510.

Grant, R. (2005) *Contemporary Strategy Analysis*. 5th edn. Blackwell, Oxford.

Green, S.D. (1996) A metaphorical analysis of client organisations and the briefing process. *Construction Management and Economics*, 14(2), 155–64.

Grey, S. (1998) *Practical Risk Assessment for Project Management*. Series in Software Engineering Practice, Wiley, Chichester.

Groote, G., Hugenholtz-Sasse, C. and Slikker, P. (2002) *Projecten Leiden* [Projects Leiden]. Het Spectrum, Utrecht (in Dutch).

Gruneberg, S.L. and Ive, G.J. (2000) *The Economics of the Modern Construction Firm*. Macmillan, Basingstoke.

Gubrium, J.F. and Holstein, J.A. (1997) *The New Language of Qualitative Method*. Oxford University Press, Oxford.

Haley, G. and Shaw, G. (2002) Is 'guaranteed maximum price' the way to go? *Hong Kong Engineer*, January.

Hall, R.H. (1972) *Organizations: Structure and Process*. Englewood Cliffs, Prentice-Hall, NJ.

Halpin, D. and Kueckmann, M. (2002) Lean construction and simulation. *Proc. 2002 Winter Simulation Conference*. San Diego, CA, 8–11 December, 1697–703.

Halpin, D. and Martinez, L. (1999) Real world applications of construction process simulation. *Proc. 1999 Winter Simulation Conference*. Phoenix, AZ, 5–8 December, 956–62.

Hancock, M.R (2000) *Cultural Differences between Construction Professionals in Denmark and United Kingdom*. SBI report 324. Danish Building Research Institute, Copenhagen.

Harty, C. (2005) Innovation in construction: a sociology of technology approach. *Building Research & Information*, 33(6), 512–22.

Hassan, M.M.D. (2006) Engineering supply chains as systems. *System Engineering*, 9(1), 73–89.

Hastrup, K. and Hervik, P. (eds) (1994) *Social Experience and Anthropological Knowledge*. Routledge, London.

Hatch, M.J. (2001). *Organisationsteori* [Organisation Theory]. Studentlitteratur, Lund.

Healey, P., Khakee, A. Motte, A. and Needham, B. (1999) European developments in strategic spatial planning. *European Planning Studies*, 7(3), 339–55.

Henecke, B. (2006) *Plan och protest, en sociologisk studie av kontroverser, demokrati och makt i den fysiska planeringen* [Plan and Protest: a Sociological Study of Controversies, Democracy and Power in the Planning Process]. Department of Sociology, Lund University, Lund.

Henecke, B. and Olander, S. (2003) *Missnöjda medborgares säkerhetsventil, en studie av överklagade detaljplaner* [The Safety Valve of Dissatisfied Citizens: a Study of Appealed Detailed Community Plans]. Departments of Construction Management and Sociology, Lund University, Lund (in Swedish).

Henig, M. and Buchanan, J. (1996) Solving MCDM problems. Process concepts. *J. Multi-Criteria Decision Analysis*, 5(1), 3–21.

Hildreth, J., Vorster, M. and Martinez, J. (2005) Reduction of short-interval GPS data for construction operation analysis. *J. Construction Engineering and Management*, 131(8), 920–7.

Hillier, B. and Hanson, J. (1984) *The Social Logic of Space*. Cambridge University Press, Cambridge.

Hillson, D.A. (2005) Describing probability: the limitations of natural language. Edinburgh. *Proc. PMI Global Congress 2005 EMEA*.

Hintze, S. (1994) *Risk Analysis in Foundation Engineering with Application to Piling in Loose Friction Soils in Urban Situations*. Doctoral thesis, Division of Soil and Rock Mechanics, Royal Institute of Technology, Stockholm.

Hintze, S., Olsson, L. and Täljsten, B. (2003) Risk och riskhantering i arbete i jord och berg [Risk and risk management in earth works and rock construction]. *Bygg och teknik* (1–2003), 12–17 (in Swedish).

Hobbs, J.E. (1996) A transaction cost approach to supply chain management. *Supply Chain Management*, 2(1), 15–27.

Hoffman, D. (2002) *Managing Operational Risk*. Wiley, New York, NY.

Holden, N.J. and von Kortzfleisch, H.F.O. (2004) Why cross-cultural knowledge transfer is a form of translation in more ways than you think. *Knowledge and Process Management*, 11(2), 127–33.

Holstein, J.A. and Gubrium, J.F. (1995) *The Active Interview*. Sage, London.

Höök, M. (2005) *Timber Volume Element Prefabrication: Production and Market Aspects*. Licentiate dissertation, 2005:65, Department of Structural Engineering, Luleå University of Technology, Luleå.

Howell, G., Macomber, H., Koskela, L. and Draper, J. (2004) Leadership and project management: time for a shift from Fayol to Flores. *Proc. 12th Annual Lean Construction Conference (IGLC-12)*. 2–6 August, Elsinore, International Group for Lean Construction (www.iglc.net).

Hultén, V. (2004) Statistics from Sveriges Byggindustrier (The Swedish Construction Federation) – personal communication and emails. Sveriges Byggindustrier, Stockholm.

Husserl, E. (1913/1962) *Ideas: General Introduction to Pure Phenomenology*, trans. W.R.B. Gibson. Allen & Unwin, London.

IEC (2001) *Project Risk Management – Application Guidelines*, International Standard IEC 62198:2001. International Electrotechnical Commission, Geneva.

Ingvaldsen, T., Lakka, A., Nielsen, A., Bertelsen, N.H. and Jonsson, B. (2004) *Productivity Studies in Nordic Building- and Construction Industry*. Project report 377. Norwegian Building Research Institute, Oslo.

IPMA (2006) *ICB-IPMA Competence Baseline Version 3.0*. International Project Management Association, Nijkerk, the Netherlands.

ISO (2000) *Quality Management Systems – Requirements (ISO 9001:2000)*. International Organization for Standardization, Geneva.

ISO (2003) *Quality Management Systems – Guidelines for Quality Management in Projects (ISO 10006:2003)*. International Organisation for Standardization, Geneva.

ISO (2004) *Environmental Management Systems – Requirements with Guidance for Use (ISO 14001:2004)*. International Organization for Standardization, Geneva.

ISO (2008) *Buildings and Constructed Assets – Service Life Planning. Part 5: Whole Life Cycle Costing (ISO/DIS15686–5)*. International Organization for Standardization, Geneva.

IT Bygg och Fastighet (2002) Online, available at: www.itbof.com/2002/ITBOF2002.html.

Jaafari, A. (2001) Management of risks, uncertainties and opportunities on projects: time for a fundamental shift. *International Journal of Project Management*, 19(2), 89–101.

Jacobsen, D.I. and Thorsvik, J. (2006) *Hur morderna organisationer fungerar* [How Modern Organizations Work]. Studentlitteratur, Lund.

Jiang, Y. (2003) The effects of traffic flow rates at freeway work zones on asphalt pavement construction productivity. *Transportation Quarterly*, 57(3), 83–103.

Johansson, B. and Svedinger, B. (1997) *Kompetensutveckling inom samhällsbyggnad: byggherren i fokus* [Competence Development in the Social Building Sector: the Construction Client in Focus]. Kungliga Ingenjörsvetenskapsakademien [Royal Academy of Engineering Sciences], Stockholm (in Swedish).

Johansson, C.E. (2003) *Visioner och verkligheter: kommunikationen om foretagets strategi* [Visions and Realities: the Communication of Company Strategy]. Doctoral thesis, Uppsala University, Uppsala (in Swedish).

Johnson, D.W. and Johnson, F.P. (2006) *Joining Together Group Theory and Group Skills*. 9th edn. Pearson Education, Boston, MA.

Johnson, G., Scholes, K. and Whittington, R. (2005) *Exploring Corporate Strategy: Text and Cases*. 7th edn. Financial Times/Prentice-Hall, Harlow.

Jonassen, D.H. and Kwon, H.I. (2001) Communication patterns in computer mediated versus face-to-face group problem solving. *Educational Technology Research and Development*, 49(1), 35–51.

Josephson, P.-E. (1994) *Orsaker till fel i byggandet: en studie om felorsaker, felkonsekvenser, samt hinder för inlärning i byggprojekt* [Causes of Defects in Building: a Study of Causes and Consequences of Defects and Barriers to Learning in Building Projects]. Doctoral thesis, Department of Construction Management, Chalmers University of Technology, Gothenburg (in Swedish).

Junnonen, J.M. (1998) Strategy formation in construction firms. *Engineering Construction and Architectural Management*, 5(2), 107–14.

Kadefors, A. (1997) *Beställar-entreprenörrelationer i byggandet: samarbete, konflikt och social påverkan* [Client-Contractor Relationships in Building Projects: Co-operation, Conflict and Social Influence]. Doctoral thesis, Department of Construction Management, Chalmers University of Technology, Gothenburg (in Swedish).

Kähkönen, K. (2001) Integration of risk and opportunity thinking in projects. *PMI Europe 2001 – Proc. 4th European Project Management Conference*, 6–7 June, London (on CD-ROM).

Kähkönen, K. and Huovila, P. (1995) Risk Management of Construction Projects in Russia. *Proc. IPMA Symposium*, St. Petersburg, 14–16 September, The Russian Project Management Association (SOVNET), 237–41.

Kalbro, T. (2002) *Rättsliga begränsningar för exploateringsavtal – en sammanfattning* [Legal Limitations for Development Agreements – a Summary]. School of Architecture and the Built Environment, Royal Institute of Technology, Stockholm (in Swedish).

Kaplan, R.S. and Norton, D.P. (2001) *The Strategy-focused Organization: how Balanced Scorecard Companies thrive in the New Business Environment*. Harvard Business School, Boston, MA.

Kartam, N. and Flood, I. (2000) Construction simulation using parallel computing environments. *Automation in Construction*, 10(1), 69–78.

Katz, D. and Kahn, R.L. (1978) *The Social Psychology of Organizations*. 2nd edn. Wiley, New York, NY.

Keast, R. and Hampson, K. (2007) Building constructive innovation networks: role of relationship management. *J. Construction Engineering and Management*, 133(5), 364–73.

Keeney, R. and Raiffa, H. (1976) *Decisions with Multiple Objectives*. Wiley, New York, NY.

Khakee, A. and Barbanente, A. (2003) Negotiative land-use and deliberative environmental planning in Italy and Sweden. *International Planning Studies*, 8(3), 181–200.

Kingdon, J.W. (1995) *Agendas, Alternatives, and Public Policies*. Harper Collins, New York, NY.

Kishk, M., Al-Hajj, A., Pollock, R., Aouad, G., Bakis, N. and Sun, M. (2003) *Whole Life Costing in Construction – a State of the Art Review*. Research paper series, 4(18). RICS Foundation, London.

Kiviniemi, A. (2006) Ten years of IFC development – why are we not yet there? CIB-W78 Keynote lecture. In *Proc. Joint International Conference on Computing and Decision Making in Civil and Building Engineering*, 14–16 June, Montreal.

Kjølle, K.H., Blakstad, S.H. and Haugen, T.I. (2005) Boundary objects for design of knowledge workplaces. *Designing Value: New Directions in Architectural Management, Proc. CIB W096 Architectural Management Conference*, 3–5 November, Copenhagen.

Klein Woolthuis, R., Hillebrand, B. and Nooteboom, B. (2005) Trust, contract and relationship development. *Organization Studies*, 26(6), 813–40.

Knauseder, I. (2007) *Organisational Learning Capabilities in Swedish Construction Projects*. Doctoral thesis, Department of Civil and Environmental Engineering, Chalmers University of Technology, Gothenburg.

Kock, N.F., McQueen, R.J. and Scott, J.L. (2000) Can action research be made more rigorous in a positivist sense? The contribution of an iterative approach. *Action Research E-Reports*, 9, September. Online, available at: www2.fhs.usyd. edu.au/arow/arer/009.htm.

Kolb, D.A. (1984) *Experiential Learning: Experience as the Source of Learning and Development*. Prentice-Hall, Englewood Cliffs, NJ.

Koskela, L. (2000) *An Exploration towards a Production Theory and its Application to Construction*. VTT Publication 408, VTT, Espoo, Finland.

Koskela, L. and Howell, G. (2002a) The underlying theory of project management is obsolete. In D.P. Slevin, J.K. Pinto and D.I. Cleland (eds) *Proc. PMI Research Conference 2002*, Project Management Institute, Newtown Square, PA, 293–302.

Koskela, L. and Howell, G. (2002b) The theory of project management: explanation to novel methods. In C.T. Formoso and G. Ballard (eds) *Proc. 10th Annual Lean Construction Conference (IGLC-10)*, 6–8 August, Gramado, International Group for Lean Construction, 1–11 (www.iglc.net).

Koskinen, K.U., Pihlanto, P. and Vanharanta, H. (2003) Tacit knowledge acquisition and sharing in a project work context. *International Journal of Project Management*, 21(4), 281–90.

Kotaji, S., Schuurmans, A. and Edwards, S. (2003) *Life Cycle Assessment in Building and Construction: a State-of-the-Art Report*. Society of Environmental Toxicology and Chemistry, Denver.

Kotter, J. (1995) Leading change: why transformation efforts fail. *Harvard Business Review*, 73(2), 59–67.

Kotter, J.P. and Heskett, J.L. (1992) *Corporate Culture and Performance*. Free Press, New York, NY.

Kramer, R.M. (1999) Trust and distrust in organizations: emerging perspectives, enduring questions. *Annual Review of Psychology*, 50, 569–98.

Kraut, R.E., Fish, R.S., Root, R.W. and Chalfonte, B.L. (1990) Informal communication in organizations: form, function, and technology. In S. Oskamp and S. Spacapan (eds) *Human Reactions to Technology: The Claremont Symposium on Applied Social Psychology*. Sage Publications, Beverly Hills, CA, 145–99.

Krishnamurthy, B.K., Tserng, H.-P., Schmitt, R.L., Russell, J.S., Bahia, H.U. and Hanna, A.S. (1998) AutoPave: towards an automated paving system for asphalt pavement compaction operations. *Automation in Construction*, 8(2), 165–80.

Kristoffersen, S. and Ljungberg, F. (1999) Making place to make IT work: empirical explorations of HCI for mobile CSCW. In *Proc. International ACM SIG-GROUP Conference on Supporting Group Work*, Phoenix, 276–85.

Kuhn, T. (1996) *The Structure of Scientific Revolutions*. 3rd edn. University of Chicago Press, Chicago, IL.

Kvale, S. (2001) *Det kvalitative forskningsintervju* [The Qualitative Research Interview]. Gyldendahl Norsk Forlag, Oslo (in Norwegian).

Lane, C. and Bachmann, R. (1997) Co-operation in interfirm relations in Britain and Germany: the role of social institutions. *British Journal of Sociology*, 48(2), 226–54.

Lansley, P. (1994) Analysing construction organizations. *Construction Management and Economics*, 12(4), 337–48.

Larsson, G. (1997) *Land Management, Public Policy, Control and Participation*. Swedish Council for Building Research [Byggforskningsrådet], Stockholm.

Latham, M. (1994) *Constructing the Team*. HMSO, London.

Latour, B. (1991) Technology is society made durable. In J. Law (ed.) *A Sociology of Monsters: Essays on Power, Technology and Domination*. Routledge, London.

Lau, E. and Rowlinson, S. (2005) The value base of trust for the construction industry. *J. Construction Procurement*, 11(1), 19–39.

Lawrence, P. and Lorsch, J. (1967) *Organization and Environment: Managing Differentiation and Integration*. Graduate School of Business Administration, Harvard University, Boston.

Leech, D. and Powell, W.D. (1974) *Levels of Compaction of Dense Coated Macadam Achieved during Pavement Construction*. Laboratory report 619. Transport Research Laboratory, Crowthorne, Berkshire.

Lemley, J.K. (1996) Image versus reality – Channel Tunnel image management. *Proc. Institution of Civil Engineers, Civil Engineering*, 114, 12–17.

Leonard-Barton, D. (1988) Implementation as mutual adaptation of technology and organization. *Research Policy*, 17(5), 251–67.

Leung, H.M., Chuah, K.B. and Tummala, V.M.R. (1998) A knowledge-based system for identifying potential project risks. *Omega – International Journal of Management Science*, 26(5), 623–38.

Lewin, K. (1946/48) Action research and minority problems. In G.W. Lewin (ed.) *Resolving Social Conflicts*. Harper & Row, New York, NY.

Li, B., Akintoye, A., Edwards, P.J. and Hardcastle, C. (2005) The allocation of risk in PPP/PFI construction projects in the UK. *International Journal of Project Management*, 23(1), 25–35.

Li, C.C., Oloufa, A.A. and Thomas, H.R. (1996) A GIS-based system for tracking pavement compaction. *Automation in Construction*, 5(1), 51–9.

Lim, G., Ahn, H. and Lee, H. (2005) Formulating strategies for stakeholder management; a case-based reasoning approach. *Expert Systems with Applications*, 28(4), 831–40.

Lindenberg, S. (2000) It takes both trust and lack of mistrust: the working of cooperation and relational signaling in contractual relationships. *J. Management and Governance*, 4(1–2), 11–33.

Ling, F., Hartmann, A., Kumaraswamy, M. and Dulaimi, M. (2007) Influences on innovation benefits during implementation: client's perspective. *J. Construction Engineering and Management*, 133(4), 306–15.

Lingard, H., Brown, K., Bradley, L., Bailey, C. and Townsend, K. (2007) Improving employees' work-life balance in the construction industry: project alliance case study. *J. Construction Engineering and Management*, 133(10), 807–15.

Linn, B. (1998) *Arkitektur som kunskap* [Architecture as Knowledge]. T10:1998, Byggforskningsrådet [Swedish Council for Building Research], Stockholm (in Swedish).

Ljung, B. (1998) *Tankar om ledning och utveckling i fastighetsföretag* [Thoughts about Management and Development of Real Estate Companies]. Byggforskningsrådet [Swedish Council for Building Research], Stockholm (in Swedish).

Locke, E.A. and Latham, G.P. (1984) *Goal Setting: A Motivational Technique That Works*. Prentice-Hall, Englewood Cliffs, NJ.

Löfgren, A. (2006) *Mobile Computing and Project Communication – Mixing Oil and Water?* Licentiate dissertation, Department of Industrial Economics and Management, Royal Institute of Technology, Stockholm.

Long, C.P. and Sitkin, S.B. (2006) Trust in the balance: how managers integrate trust-building and task-control. In R. Bachmann and A. Zaheer (eds) *Handbook of Trust Research*. Edward Elgar, Cheltenham, 87–106.

Loosemore, M. and Tan, C.C. (2000) Occupational stereotypes in the construction industry. *Construction Management and Economics*, 18(18), 559–66.

Lousberg, L.H. (2006) Towards a theory of project management. In D. Amaratunga *et al.* (eds) *Proc. 6th International Postgraduate Research Conference*. 6–8 April, University of Salford and Delft University of Technology, Delft, the Netherlands, 40–53.

Love, P.E.D., Irani, Z., Cheng, E. and Li, H. (2002) A model for supporting inter-organizational relations in the supply chain. *Engineering, Construction and Architectural Management*, 1(9), 2–15.

Love, P.E.D., Li, H., Irani, Z. and Faniran, O. (2000) Total quality management and the learning organization: a dialogue for change in construction. *Construction Management and Economics*, 18(3), 321–31.

Lowe, J. (1987) Monopoly and the material supply industries of the UK. *Construction Management and Economics*, 5(1), 57–71.

Luhmann, N. (1993) *Risk: a Sociological Theory*. Walter de Gruyter, Berlin.

Lyons, T. and Skitmore, M. (2004) Project risk management in the Queensland engineering construction industry: a survey. *International Journal of Project Management*, 22(1), 51–61.

McAllister, D.J. (1995) Affect- and cognition-based trust as foundations for interpersonal cooperation in organizations. *Academy of Management Journal*, 38(1), 24–59.

Macaulay, S. (1963) Non-contractual relations in business: a preliminary study. *American Sociological Review*, 28(1), 55–67.

McCloughan, P. (2004) Construction sector concentration: evidence from Britain. *Construction Management and Economics*, 22(9), 979–90.

McEvily, B., Perrone, V. and Zaheer, A. (2003), Trust as an organizing principle. *Organization Science*, 14(1), 91–103.

Madhok, A. (2006) Opportunism, trust and knowledge: the management of firm value and the value of firm management. In R. Bachmann and A. Zaheer (eds) *Handbook of Trust Research*. Edward Elgar, Cheltenham, 107–23.

Mahesh, G., Kumaraswamy, M., Anvuur, A. and Coffey, V. (2007) Contracting for community development: a case study based perspective of a public sector client initiative in Hong Kong. In S.M. Ahmed, S. Azhar and S. Mohamed (eds) *Fourth International Conference on Construction in the 21st Century (CITC-IV) – Accelerating Innovation in Engineering, Management and Technology*. Gold Coast, Australia, 366–72.

Malhotra, D. and Murninghan, J.K. (2002) The effects of contracts on interpersonal trust. *Administrative Science Quarterly*, 47(3), 534–59.

Malterud, K. (1995) Action research – a strategy for evaluation of medical interventions. *Family Practice*, 12(4), 476–81.

March, J.G. (1994) *A Primer on Decision Making: How Decisions Happen*. Free Press, New York, NY.

March, J.G. (1998) *Decisions and Organizations*. Blackwell, Oxford.

March, J.G. and Shapira, Z. (1987) Managerial perspectives on risk and risk taking. *Management Science*, 33(11), 1404–18.

March, J.G. and Simon, H.A. (1993) *Organizations*. 2nd edn. Blackwell, Oxford.

Marcotte, C. and Niosi, J. (2000) Technology transfer to China: the issues of knowledge and learning. *J. Technology Transfer*, 25(1), 43–57.

Mayer, R.C., Davis, J.H. and Schoorman, F.D. (1995) An integrative model of organizational trust. *Academy of Management Review*, 20(3), 709–34.

Mayo, E. (1945) *The Social Problems of an Industrial Civilization*. Graduate School of Business Administration, Harvard University, Boston, MA.

Merriam, S.B. (1998) *Case Qualitative Research and Case Study Applications*. Jossey-Bass, San Francisco, CA.

Meyer, J.W. (1996) Otherhood: the promulgation and transmission of ideas in the modern organizational environment. In B. Czarniawska and G. Sevón (eds) *Translating Organizational Change*. Walter de Gruyter, Berlin, 241–52.

Midgley, J. (2000) Exploring alternative methodologies to establish the effects of land area designation in development control decisions. *Planning Practice & Research*, 15(4), 319–33.

Milgrom, P. and Roberts, J. (1992) *Economics, Organization, and Management*. Prentice-Hall, Englewood Cliffs, NJ.

Miller, T. (1993) *Genomföandeavtal i exploateringsprocessen* [Public Implementation Agreements in the Real Estate Development Process]. Boverket [National Board of Housing, Building and Planning], Karlskrona (in Swedish).

Miniace, J.N. and Falter, E. (1996) Communication: a key factor in strategy implementation. *Planning Review*, 24(1), 26–30.

Mintzberg, H. (1976) The structure of 'unstructured' decision processes. *Administrative Science Quarterly*, 21(2), 246–75.

Mintzberg, H. (1979) *The Structuring of Organizations: a Synthesis of the Research*. Prentice-Hall, Englewood Cliffs, NJ.

Mintzberg, H. (1983) *Structure in Fives. Designing Effective Organizations*. Prentice-Hall, Englewood Cliffs, NJ.

Mintzberg, H., Ahlstrand, B.W. and Lampel, J. (1998) *Strategy Safari: a Guided Tour through the Wilds of Strategic Management*. Free Press, New York, NY.

Mintzberg, H., Raisinghani, D. and Théorêt, A. (1976) The structure of 'unstructured' decision processes. *Administrative Science Quarterly*, 21(2), 246–75.

Mitchell, R.K., Bradley, R.A. and Wood, D.J. (1997) Toward a theory of stakeholder identification and salience: defining the principle of who and what really counts. *Academy of Management Review*, 22(4), 853–85.

Moles, P. and Terry, N. (1997) *The Handbook of International Financial Terms*. Oxford University Press, Oxford.

Monk, S. and Whitehead, C.M.E. (1999) Evaluating the economic impact of planning controls in the United Kingdom: some implications for housing. *Land Economics*, 75(1), 74–94.

Montgomery, D.C. (2005) *Statistical Quality Control*. Wiley, New York, NY.

Morecroft, J.D.W. (1988) Systems dynamics and microworlds for policymakers. *European J. Operational Research*, 35(3), 301–20.

Morecroft, J.D.W. (1992) Executive knowledge, models and learning. *European Journal of Operational Research*, 59(1), 9–27.

Morris, P. (1972) *A Study of Selected Building Projects in the Context of Theories of Organization*. Doctoral thesis, Department of Building, UMIST, Manchester.

Motawa, I.A., Anumba, C.J. and El-Hamalawi, A. (2006) A fuzzy system for evaluating the risk of change in construction projects. *Advances in Engineering Software*, 37(9), 583–91.

Mouton, J. (2001) *How to Succeed in your Master's & Doctoral Studies – A South African Guide and Resource Book*. Van Schaik Publishers, Pretoria.

Müllern, T. and Stein, J. (1999) *Övertygandets ledarskap: om retorik vid strategiska förändringar* [Persuasive Leadership: about Rhetoric in Strategic Change]. Studentlitteratur, Lund (in Swedish).

Nahapiet, H. and Nahapiet, J. (1985) A comparison of contractual arrangements for building projects. *Construction Management and Economics*, 3(2), 217–31.

Nam, C.B. and Tatum, C.B. (1992) Non-contractual methods of integration on construction projects. *J. Construction Engineering and Management*, 118(2), 385–98.

Naresh, A.L. and Jahren, C.T. (1999) Learning outcomes from construction simulation modeling. *Civil Engineering and Environmental Systems*, 16(2), 129–44.

Nassar, K.M., Nassar, W.M. and Hegab, M.Y. (2005) Evaluating cost overruns of asphalt paving project using statistical process control methods. *J. Construction Engineering and Management*, 131(11), 1173–8.

Nässén, J. and Holmberg, J. (2005) Energy efficiency – a forgotten goal in the Swedish building sector? *Energy Policy*, 33(8), 1037–51.

Navon, R., Goldschmidt, E. and Shpatnisky, Y. (2004) A concept proving prototype of automated earthmoving control. *Automation in Construction*, 13(2), 225–39.

New, S. and Westbrook, R. (eds) (2004) *Understanding Supply Chains: Concepts, Critiques and Futures*. Oxford University Press, Oxford.

Nicholas, J.M. (2004) *Project Management for Business and Engineering*. Elsevier Butterworth-Heinemann, Burlington, MA.

Nielsen, J. (1993) *Usability Engineering*. Morgan Kaufmann, San Francisco, CA.

Nilsson, A. and Hellström, D. (2001) *Miljöbarometern 2001* [The Swedish Environmental Barometer 2001]. Internal report of the Swedish Business Environmental Barometer 2001, HandelsConsulting, Gothenburg (in Swedish).

Nonaka, I. and Nishiguchi, T. (2001) Knowledge emergence. In I. Nonaka and T. Nishiguchi (eds) *Knowledge Emergence: Social, Technical, and Evolutionary Dimensions of Knowledge Creation*. Oxford University Press, Oxford, 3–10.

Nonaka, I. and Takeuchi, H. (1995) *The Knowledge-creating Company*. Oxford University Press, Oxford.

Noorderhaven, N.G. (1995) *Strategic Decision Making*. Addison-Wesley, Wokingham, Berks.

Nooteboom, B. (2002) *Trust: Forms, Foundations, Functions, Failures and Figures*. Edward Elgar, Cheltenham.

NS (2000) *NS3454 Life Cycle Costs for Building and Civil Engineering Work – Principles and Classification*. Standard Norge, Oslo.

Nutt, P.C. (1993) The formulation processes and tactics used in organizational decision-making. *Organization Science*, 4(2), 226–51.

Nutt, P.C. (2001) A taxonomy of strategic decisions and tactics for uncovering alternatives. *European Journal of Operational Research*, 132(3), 505–27.

Nutt, P.C. (2004) Averting decision debacles. *Technological Forecasting and Social Change*, 71(3), 239–65.

Nutt, P.C. (2005) Search during decision making. *European Journal of Operational Research*, 160(3), 851–76.

Öberg, M. (2005) *Integrated Life Cycle Design – Application to Swedish Concrete Multi-dwelling Buildings*. Doctoral thesis, Department of Building Materials, Lund University, Lund.

Olander, S. (2006) *External Stakeholder Analysis in Construction Project*

Management. Doctoral thesis, Department of Construction Management, Lund University, Lund.

Olander, S. (2007) Stakeholder impact analysis in construction project management. *Construction Management and Economics*, **25**(3), 277–87.

Olander, S. and Landin, A. (2005) Evaluation of stakeholder influence in the implementation of construction projects. *International Journal of Project Management*, **23**(4), 321–8.

Olander, S. and Landin, A. (2008) Housing developers' perceptions of the planning process: a survey of Swedish companies. *International Journal of Housing Markets and Analysis*, **1**(3), 246–55.

Oliver, C. (1991) Strategic responses to institutional processes. *Academy of Management Review*, **16**(1), 145–79.

Oloufa, A.A. (2002) Quality control of asphalt compaction using GPS-based system architecture. *IEEE Robotics & Automation Magazine*, **9**(1), 29–35.

Osipova, E. and Apleberger, L. (2007) Risk management in different forms of contract and collaboration – case of Sweden. *Proc. CIB World Building Congress: Construction for Development*. 14–17 May, Cape Town, CIB, Rotterdam, the Netherlands.

Ouchi, W. (1980) Markets, bureaucracies and clans. *Administrative Science Quarterly*, **25**(1), 129–41.

Öztas, A. and Ökmen, O. (2005) Judgmental risk analysis process development in construction projects. *Building and Environment*, **40**(9), 1244–54.

Palmer, A. and McGeorge, D. (1997) The relationship of stakeholders to soft system construction management techniques. In C.H. Davidson and T.A.A. Meguid (eds) *Procurement – a Key to Innovation. Proc. CIB W92 Symposium on Procurement*. University of Montreal.

Pampagnin, L.-H., Peyret, F. and Garcia, G. (1998) Architecture of a GPS-based guiding system for road compaction. *Robotics and Automation. Proc. 1998 IEEE International Conference*, Leuven, 3, 2422–7.

Peek, G.J. (2006) *Locatiesynergie: een participatieve start van de herontwikkeling van binnenstedelijke stationslocaties* [Location Synergy: a Participative Start for the Redevelopment of Inner City Station Areas]. Eburon Academic Publishers, Delft (in Dutch).

Peiser, R.B. and Frej, A.B. (2003) *Professional Real Estate Development: The ULI Guide to the Business*. 2nd edn. Urban Land Institute, Washington, DC.

Persson, M. (1999) Ny byggprocess – Svedalamodellen [New construction process – the Svedala model]. In J. Bröchner and P.-E. Josephson (eds) *Construction Economics and Organisation, Proc. Nordic Seminar on Construction Economics and Organization*. 12–13 April, Department of Construction Management, Chalmers University of Technology, Gothenburg, 45–52.

Persson, M. (2006) *Lessons Learned on Knowledge Management – The Case of Construction*. Doctoral thesis, Department of Construction Sciences, Lund University, Lund.

Persson, M., Landin, A. and Andersson, A. (2006) *Assembling Knowledge and Sharing Experience within a Contracting Company – a Case Study*. Report LUTVDG/TVBP-06/3088-SE. Department of Construction Sciences, Lund University, Lund.

Peyret, F., Jurasz, J., Carrel, A., Zekri, E. and Gorham, B. (2000) The computer integrated road construction project. *Automation in Construction*, **9**(5), 447–61.

Phillips, R. (2003) *Stakeholder Theory and Organizational Ethics*. Berrett-Koehler Publishers, San Francisco, CA.

Phua, F.T.T. and Rowlinson, S. (2003) Cultural differences as an explanatory variable for adversarial attitudes in the construction industry: the case of Hong Kong. *Construction Management and Economics*, **21**(7), 777–85.

Pijnacker Hordijk, E.H., van der Bend, G.W. and van Nouhuys, J.F. (2004) *Aanbestedingsrecht – Handboek voor het Europese en het Nederlandse Aanbestedingsrecht* [Procurement Law – Manual of European and Dutch Procurement Laws]. Sdu Publishers, The Hague, 366.

PMI (2004) *A Guide to the Project Management Body of Knowledge*. 3rd edn. PMBOK Guides. Project Management Institute, Newtown Square, PA.

Polanyi, M. (1967) *The Tacit Dimension*. Anchor Books, New York, NY.

Poppo, L. and Zenger, T. (2002) Do formal contracts and relational governance function as substitutes of complements? *Strategic Management Journal*, **23**(8), 707–25.

Porter, M.E. (1985) *Competitive advantage*. Free Press, New York, NY.

Post, J.E., Preston, L.E. and Sachs, S. (2002*) Redefining the Corporation – Stakeholder Management and Organizational Wealth*. Stanford University Press, Stanford, CA.

Powell, W.W. and Dimaggio, P.J. (eds) (1991) *The New Institutionalism in Organizational Analysis*, University of Chicago Press, Chicago, IL.

Prahalad, C.K. and Hamel, G. (1990) The core competence of the corporation. *Harvard Business Review*, **68**(3), 79–91.

Prencipe, A. and Tell, F. (2001) Inter-project learning: processes and outcomes of knowledge codification in project-based firms. *Research Policy*, **30**(9), 1373–94.

Price, A.D.F. (2003) The strategy process within large construction organisations. *Engineering, Construction and Architectural Management*, **10**(4), 283–96.

Projectorganisatie Stationsgebied (2003) *Masterplan Stationsgebied Utrecht* [Utrecht Station Master Plan]. Gemeente Utrecht [Commune of Utrecht] (in Dutch).

Pryke, S. and Smyth, H. (2006) *The Management of Complex Projects*. Blackwell, Oxford.

PSIB (2003) *Proces en Systeem Innovatie in de Bouw* [Rethinking the Dutch Construction Industry]. PSIB Programmabureau, Gouda (in Dutch).

Raftery, J. (1994) *Risk Analysis in Project Management*. Spon, London.

Rahman, M. and Kumaraswamy, M. (2004) Contracting relationship trends and transitions. *J. Management in Engineering*, **20**(4), 147–61.

Reve, T. and Levitt, R.E. (1984) Organisation and governance in construction. *Project Management*, **2**(1), 17–25.

Rigby, C., Day, M., Forrester, P. and Burnett, J. (2000) Agile supply: rethinking systems thinking, systems practice. *International Journal of Agile Management Systems*, **2**(3), 178–86.

Riksdagens revisorer [Parliamentary auditors] (2001) *Plan- och byggprocessens längd* [The Length of the Planning and Building Process]. Report 2000/01:14, Swedish Parliament, Stockholm (in Swedish).

Ring, P.S. and Ven de Ven, A.H. (1994) Developmental processes of cooperative interorganizational relationships. *Academy of Management Review*, **19**(1), 90–118.

Rittel, H.W.J. and Webber, M.M. (1973) Dilemmas in a general theory of planning. *Policy Science*, **4**(2), 155–69.

Roberts, F.L, Kandhal, P.S., Brown, E.R., Lee, D.Y. and Kennedy, T.W. (1991) *Hot Mix Asphalt Materials, Mixture Design, and Construction*. NAPA Education Foundation, Lanham, MD.

Robins, J. (1987) Organizational economics: notes on the use of transaction theory in the study of organizations. *Administrative Science Quarterly*, **32**(1), 68–86.

Robson, C. (2002) *Real World Research: a Resource for Social Scientists and Practitioner-Researchers*. 2nd edn. Blackwell, Oxford.

Roelofs, B. and Reinderink, H. (2005) *Bouworganisatie- en contractvormen – ordening, standaardisering en toepassing* [Organisation and Contract Forms in the Construction Industry – Structuring, Standardisation and Application]. ONRI, The Hague (in Dutch).

Rogers, E.M. (2003) *Diffusion of Innovations*. 5th edn. Free Press, New York, NY.

Rooke, J., Seymour, D. and Fellows, R. (2004) Planning for claims: an ethnography of industry culture. *Construction Management and Economics*, **22**(6), 655–62.

Rorty, R. (1967) *The Linguistic Turn: Recent Essays in Philosophical Method*. University of Chicago Press, Chicago, IL.

Rosenberg, N. (1982) Learning by using. In N. Rosenberg (ed.) *Inside the Black Box: Technology and Economics*. Cambridge University Press, Cambridge, 120–40.

Rosenhead, J. and Mingers, J. (2001) Rational analysis for a problematic world revisited. In *Problem Structuring Methods for Complexity, Uncertainty and Conflict*. Wiley, Chichester.

Rouse, W.B. (2005) Enterprises as systems: essential challenges and approaches to transformation. *System Engineering*, **8**(2), 138–50.

Rousseau, D., Sitkin, S., Burt, R. and Camerer, C. (1998) Not so different after all: a cross discipline view of trust. *Academy of Management Review*, **23**(3), 393–404.

Rowlinson, S. (2008) Corporate social responsibility in the Hong Kong and Asia Pacific construction. In M. Murray and A.J.R. Dainty (eds) *Corporate Social Responsibility in the Construction Industry*. Taylor and Francis, London, 327–50.

Rowlinson, S., Cheung, F.Y.K., Simons, R. and Rafferty, A. (2006) Alliancing in Australia – no litigation contracts: a tautology? *ASCE Journal of Professional Issues in Engineering Education and Practice: Special Issue on Legal Aspects of Relational Contracting*, **132**(1), 77–81.

Rubino, B. (2006a) Challenging problems for a new generation of demonstration projects. In D. Amaratunga *et al.* (eds) *Proc. 6th International Postgraduate Research Conference*. 6–8 April, University of Salford and Delft University of Technology, Delft, the Netherlands.

Rubino, B. (2006b) *Pilotprojektet Hamnhuset: dokumentation och uppföljning av projekteringsprocessen* [The Hamnhuset Pilot Project: Documentation and Following-up of the Design Process]. Boverket/Chalmers University of Technology, Gothenburg (in Swedish).

Russo, J.E. and Schoemaker, P.H. (2002) *Winning Decisions: Getting it Right the First Time*. Random House, New York, NY.

Ryd, N. (2001) *Byggnadsprogram som informationsbärare i byggprocessen* [Briefing as Carrier of Information in the Building Process]. Licentiate dissertation, Department of Industrial Planning, Chalmers University of Technology, Gothenburg (in Swedish).

Ryd, N. (2003) *Exploring Construction Briefing: from Document to Process*. Doctoral thesis. Department of Space and Processes, Chalmers University of Technology, Gothenburg.

Saad, M., Jones, M. and James, P. (2002) A review of the progress towards the adoption of supply chain management (SCM) relationships in construction. *European Journal of Purchasing and Supply Management*, **8**, 179–83.

Saaty, T.L. (1980) *The Analytic Hierarchy Process*. McGraw-Hill, New York, NY.

Sahlin-Andersson, K. (1986). *Beslutsprocessens komplexitet: att genomföra och hindra stora projekt* [The Complexity of Decision Processes: to Implement and Obstruct Major Projects]. Doctoral thesis. Umeå University, Umeå.

Samuelson, O. (2003) *IT-användning i byggande och förvaltning* [IT Use in Construction and Facility Management]. Licentiate dissertation, Department of Industrial Economics and Management, Royal Institute of Technology, Stockholm (in Swedish).

Sarja, A. (2002) *Integrated Life Cycle Design of Structures*. Spon, London.

Sawhney, A., AbouRizk, S.M. and Halpin, D.W. (1998) Construction project simulation using CYCLONE. *Canadian Journal of Civil Engineering*, **25**(1), 16–25.

Scarbrough, H., Swan, J., Laurent, S., Bresnen, M., Edelman, L. and Newell, S. (2004) Project-Based Learning and the Role of Learning Boundaries. *Organization Studies*, **25**(9), 1579–600.

Schein, E.H. (1985) *Organisational Culture and Leadership: A Dynamic View*. Jossey-Bass, San Francisco, CA.

Schmenner, R.W. (1993) *Production/Operations Management: From the Inside Out*. Macmillan, New York, NY.

Schön, D.A. (1987) *Educating the Reflective Practitioner: Toward a New Design for Teaching and Learning in the Professions*. Jossey-Bass, San Francisco, CA.

Senge, P.M. (1990) *The Fifth Discipline: The Art & Practice of the Learning Organisation*. Doubleday, New York.

Senge, P.M., Roberts, C., Ross, R.B., Smith, B.J. and Kleiner, A. (1994) *The Fifth Discipline Fieldbook – Strategies and Tools for Building a Learning Organization*. Nicholas Brealey Publishing, London.

Shirazi, B., Langford, D. and Rowlinson, S. (1996) Organizational structures in the construction industry. *Construction Management and Economics*, **14**(3), 199–212.

Sijpersma, R. and Buur, A.P. (2005) *Bouworganisatievorm in beweging* [Construction Organisation on the Move]. Economisch Instituut voor de Bouwnijverheid, Amsterdam (in Dutch).

Silva, A.S. (2002) Indecision factors when planning for land use change. *European Planning Studies*, **10**(3), 335–58.

Simon, H.A. (1957) *Administrative Behavior: a Study of Decision-making Processes in Administrative Organisations*. 2nd edn. Free Press, New York, NY.

Simon, H.A. (1977) *The New Science of Management Decision*. Prentice-Hall, Englewood Cliffs, NJ.

Simon, M., Houghton, S.M. and Aquino, K. (2000) Cognitive biases, risk perception, and venture formation: how individuals decide to start companies. *J. Business Venturing*, **15**(2), 113–34.

Simons, B.J.A.G. (2006) *Op weg naar een beheerst asfaltverwerkingsproces* [Towards a More Controlled Asphalt Paving Process]. University of Twente, Enschede (in Dutch).

Sitkin, S.B. (1995) On the positive effect of legalization on trust. In R.J. Bies, R.J. Lewicki and B.H. Sheppard (eds) *Research on Negotiations in Organizations*, **5**, 185–217.

Sitkin, S.B. and Pablo, A.L. (1992) Reconceptualizing the determinants of risk behavior. *Academy of Management Review*, **17**(1), 9–38.

Slaughter, E.S. (1998) Models of construction innovation. *J. Construction Engineering and Management*, **124**(3), 226–31.

Smith, N.J. (1999) *Managing Risk in Construction Projects*. Blackwell, Oxford.

Smullen, J. and Hand, N. (2005) *A Dictionary of Finance and Banking*. Oxford University Press, Oxford.

Söderberg, J. (1994) Byggprocessen nu och i framtiden [The construction process now and in the future], from the anthology: *Från nyproduktion till fastighetsföretagande* [From New Production to Facilities Management]. T3:1994, Byggforskningsrådet [Swedish Council for Building Research], Stockholm (in Swedish).

Sohtell, B. and Sundell, U. (1993) *Genomförandeavtal – inriktning mot ekonomiska regeleringar* [Public Implementation Agreement – Emphasis on Economic Regulations]. LMV report 1993:13. Lantmäteriet, Gävle (in Swedish).

Sprei, F. (2007) *Challenges for End-use Energy Efficiency – Studies of Residential Heating and Personal Transportation in Sweden*. Licentiate dissertation, Chalmers University of Technology, Gothenburg.

Stallen, P.J.M. (2002) Risico is bias: en het kan ook niet anders [Risk is bias: what else could it be?]. *Bedrijfskunde*, **74**(3), 14–20 (in Dutch).

Star, S.L. and Griesemer, J.R. (1989) Institutional ecology, translations, and coherence: amateurs and professionals in Berkeley's Museum of Vertebrate Zoology, 1907–1939. *Social Studies of Science*, **19**(3), 387–420.

Sternberg, E. (1997) The defects of stakeholder theory. *Corporate Governance: An International Review*, **5**(1), 3–9.

Stevens, J. (1989) Integrating the supply chain. *International Journal of Physical Distribution and Materials Management*, **19**(8), 3–8.

Stigler, G. (1968) Barriers to entry, economies of scale, and firm size. In G. Stigler (ed.) *The Organization of Industry*. Richard D. Irwin, Homewood, IL.

Stinchcombe, A.L. (1959) Bureaucratic and craft administration of production. *Administrative Science Quarterly*, **4**(2), 177–95.

Stinchcombe, A.L. (1985) Contracts as hierarchical documents. In A. Stinchcombe and C. Heimer, *Organization Theory and Project Management: Administering Uncertainty in Norwegian Offshore Oil*. Norwegian University Press, Bergen, 121–71.

Stroup-Gardiner, M., Wagner, C.T., Hodgson, D.T. and Sain, J. (2002) Effect of temperature differentials on density and smoothness. In M. Stroup-Gardiner (ed.) *Constructing Smooth Hot Mix Asphalt (HMA) Pavements*. ASTM Special Technical Publication 1433, ASTM International, West Conshohocken, PA, 127–41.

Suddaby, R. (2006) From the editors: what grounded theory is not. *Academy of Management Journal*, **49**(4), 633–42.

Sullivan, P.H. (2001) *Profiting from Intellectual Capital: Extracting Value from Innovation*. Wiley, New York, NY.

Susman, G.I. (1983) Action research: a sociotechnical systems perspective. In G. Morgan (ed.) *Beyond Method: Strategies for Social Science Research*. Sage, London.

Sveiby, K.E. (1997) *The New Organizational Wealth: Managing & Measuring Knowledge-Based Assets*. Berrett-Koehler Publishers, San Francisco, CA.

Sveiby, K.E. and Risling, A. (1986) *Kunskapsföretaget – seklets viktigaste ledarutmaning?* [Knowledge Company – the Century's Most Important Leadership Challenge?] 1:5, Liber, Malmö (in Swedish).

Sverlinger, P.-O.M. (2000) *Managing Knowledge in Professional Service Organizations: Technical Consultants Serving the Construction Industry.* Doctoral thesis, Department of Service Management, Chalmers University of Technology, Gothenburg.

Swedberg, R., Himmelstrand, U. and Brulin, G. (1987) The paradigm of economic sociology. *Theory and Society,* 16(2), 169–213.

Tah, J.H.M. and Carr, V. (2000) A proposal for construction project risk assessment using fuzzy logic. *Construction Management and Economics,* 18(4), 491–500.

Tatikonda, M.V. and Rosenthal, S.R. (2000) Technology novelty, project complexity and product development project execution success: a deeper look at task uncertainty in product innovation. *IEEE Transactions on Engineering Management,* 47(1), 74–87.

Teisman, G.R. (1992) *Complexe besluitvorming: een pluricentrisch perspectief op besluitvorming over ruimtelijke investeringen* [Complex Decision Making: a Pluricentralist View of Decision-making on Spatial Investments]. Doctoral thesis, Vuga, 's-Gravenhage (in Dutch).

Teisman, G.R. (2000) Models for research into decision-making processes: on phases, streams and decision-making rounds. *Public Administration,* 78(4), 937–56.

Teisman, G.R. (2001) *Besluitvorming en ruimtelijk procesmanagement* [Decision-making and the Planning Process]. Eburon Academic Publishers, Delft (in Dutch).

Tenah, K. (1986) Management level as defined and applied within a construction organization by some US contractors and engineers. *Project Management,* 4(4), 195–204.

Thomas, R., Marosszeky, M., Karim, K., Davis, S. and McGeorge, D. (2002) The importance of project culture in achieving quality outcomes in construction. In C.T. Formoso and G. Ballard (eds) *Proc. 10th Annual Lean Construction Conference (IGLC-10).* 6–8 August, Gramado, International Group for Lean Construction, 1–11 (www.iglc.net).

Thompson, J.D. (1967) *Organizations in Action.* McGraw-Hill, New York, NY.

Tichy, N.M. (1983) *Managing Strategic Change: Technical, Political and Cultural Dynamics.* Wiley, New York, NY.

Timm, D., Voller, V.R., Lee, E. and Harvey, J. (2001) Calcool: a multi-layer asphalt pavement cooling tool for temperature prediction during construction. *The International Journal of Pavement Engineering,* 2, 169–85.

Tirole, J. (1988) *The Theory of Industrial Organization.* MIT Press, Cambridge, MA.

Trochim, W.M.K. (2001) *The Research Methods Knowledge Base.* 2nd edn. Atomic Dog Publishing Cincinnati, OH.

Turin, D.A. (2003) Building as a process. *Building Research & Information,* 31(2), 180–7.

Turner, B. (1997) Municipal housing companies in Sweden: on or off the market? *Housing Studies,* 12(4), 477–89.

Tversky, A. and Kahneman, D. (1974) Judgment under uncertainty: heuristics and biases. *Science,* 185(4157), 1124–31.

Uher, T.E. and Toakley, A.R. (1999) Risk management in the conceptual phase of a project. *International Journal of Project Management,* 17(3), 161–9.

US Census (2005) *Industry General Summary: 2002 Report Census.* October. Online, available at: www.census2010.gov/prod/ec02/ec0223sg1.pdf.

van Aken, J.E. (2005) Management research as a design science: articulating the research products of mode 2 knowledge production in management. *British Journal of Management,* 16(1), 19–36.

van Eijk, P.J. (2003) *Vernieuwen mét water: een participatieve strategie voor de gebouwde omgeving* [Refresh with Water: a Participative Strategy for the Built Environment]. Doctoral thesis, Eburon Academic Publishers, Delft (in Dutch).

van Enk, W. (2006) Top-101 Ontwikkelaars [Top 101 Real estate development companies]. *PropertyNL Magazine,* 6(17), 49–67.

Velzen, M. (2005) Building on flexibility, training for construction: a comparison of training arrangements for flexible workers in the Dutch and American construction industries. In I.U. Zeytinoglu (ed.) *Flexibility in Workplaces: Effects on Workers, Work Environment and the Unions.* IIRA/International Labour Organization, Geneva.

Verhage, R. (2002) *Local Policy for Housing Development: European Experiences.* Ashgate Publishing, Aldershot.

Vlaar, P.W.K., van den Bosch, F. and Volberda, H.W. (2007) On the evolution of Trust, distrust, and formal coordination and control in interorganizational relationships. *Group and Organization Management,* 32(4), 407–28.

von Hippel, E. (1988) Users as innovators. In E. von Hippel (ed.) *The Sources of Innovation.* Oxford University Press, Oxford, 11–27.

Vos, P.G.J.C., van Meel, J.J. and Dijcks, A.A.M. (1997) *The Office, The Whole Office and Nothing but the Office. A Framework of Workplace Concepts.* Department of Real Estate & Project Management, Delft University of Technology, Delft, the Netherlands.

Voss, C.A. (1988) Implementation: a key issue in manufacturing technology: the need for a field of study. *Research Policy,* 17(2), 55–63.

Vrijhoef, R. and de Ridder, H. (2005) Supply chain integration for achieving best value for construction clients: client-driven versus supplier-driven integration. In *Proc. QUT Research Week,* 4–6 July 2005, Brisbane.

Vrijhoef, R. and Koskela, L. (2000) The four roles of supply chain management in construction. *European Journal of Purchasing & Supply Management,* 6(3–4), 169–78.

Walker, A. (1996) *Project Management in Construction.* 3rd edn. Blackwell, Oxford.

Walker, D.H.T. and Rowlinson, S. (eds) (2008) *Procurement Systems: a Cross-industry Project Management Perspective.* Taylor and Francis, Abingdon.

Walker, D.H.T., Bourne, L. and Rowlinson, S. (2008) Stakeholders and the supply chain. In D.H.T. Walker and S. Rowlinson (eds) *Procurement System: a Cross-industry Management Perspective.* Taylor and Francis, Abingdon, 70–100.

Wamelink, J. (2006) *Inspireren, Integreren, Innoveren* [Inspiring, Integrating, Innovating]. Inaugural address, Delft University of Technology, Delft.

Ward, S. and Chapman, C. (2003) Transforming project risk management into project uncertainty management. *International Journal of Project Management,* 21(2), 97–105.

Warsame, A. (2006) *Long Run Relationships, Vertical Integration and International Competition: can they Contribute to Explaining Regional Construction*

Cost Differences? Licentiate dissertation, Department of Real Estate and Construction Management, Royal Institute of Technology, Stockholm.

Wedegren, T. (1997) *Exploateringsavtal* [Property Development Agreements]. Faculty of Law, Stockholm University, Stockholm (in Swedish).

Weick, K.E. (1995) *Sensemaking in Organizations.* Sage, Thousand Oaks, CA.

Weilenmann, A. (2003) *Doing Mobility.* Doctoral thesis, Department of Informatics. Gothenburg University, Gothenburg.

Wenger, E. (1998) *Communities of Practice: Learning, Meaning and Identity.* Cambridge University Press, New York, NY.

White, G.C., Mahoney, J.P., Turkiyyah, G.M., Willoughby, K.A. and Brown, E.R. (2002) Online tools for hot-mix asphalt monitoring. *Construction 2002*, Transportation Research Record 1813, 124–32.

Whittaker, S., Frohlich, D. and Daly-Jones, O. (1994), Informal workplace communication: what is it like and how might we support it? In *Proc. CHI '94 Conference on Human Factors in Computing Systems*, Boston, MA, 131–7.

Whittington, R. (1993) *What is Strategy – and does it Matter?* Routledge, London.

Wiedersheim-Paul, F. and Eriksson, L.T. (1989) *Att utreda och rapportera* [Investigate and Report]. Liber, Malmö (in Swedish).

Wikforss, Ö. (ed.) (2006) *Kampen om kommunikationen: om projektledningens informationsteknologi* [The Fight over Communication: about Information Technology in Project Management]. Department of Industrial Economics and Management, Research report, Royal Institute of Technology, Stockholm (in Swedish).

Williamson, O.E. (1975) *Markets and Hierarchies: Analysis and Antitrust Implication.* Free Press, New York, NY.

Williamson, O.E. (1979) Transaction-cost economics: the governance of contractual relations. *J. Law and Economics*, 22(2), 233–61.

Williamson, O.E. (1985a) *The Economic Institutions of Capitalism: Firms, Markets, Relational Contracting.* Free Press, New York, NY.

Williamson, O.E. (1985b) Reflections on the new institutional economics, *J. Institutional and Theoretical Economics*, 141(1), 187–95.

Williamson, O.E. (1991) Comparative economic organization: the analysis of discrete structural alternatives. *Administrative Science Quarterly*, 36(2), 269–96.

Williamson, O.E. (1993) Calculativeness, trust, and economic organization. *J. Law and Economics*, 36(1), 453–86.

Willoughby, K.A., Mahoney, J.P., Pierce, L.M., Uhlmeyer, J.S. and Anderson, K.W. (2002) Construction-related asphalt concrete pavement temperature and density differentials. *Construction 2002*. Transportation Research Record 1813, 68–76.

Winch, G.M. (1989) The construction firm and the construction project: a transaction cost approach. *Construction Management and Economics*, 7(4), 331–45.

Winch, G.M. (2000) Institutional reform in British construction. *Building Research & Information*, 28(1), 141–55.

Winch, G.M. (2002) *Managing Construction Projects.* Blackwell, Oxford.

Winch, G.M. (2003) Models of manufacturing construction process: the genesis of re-engineering construction. *Building Research & Information*, 31(2), 107–18.

Woodall, B. (1996) *Japan under Construction: Corruption, Politics, and Public Works.* University of California Press, Berkeley, CA.

Wortmann, J.C. (1992) Production management systems for one-of-a-kind products. *Computers in Industry*, 19(1), 79–88.

Yin, R.K. (2003) *Case Study Research: Design and Methods.* 3rd edn. Sage Publications, Thousand Oaks, CA.

Yngvesson, N., Eskiltorp, E., Lutz, J., Magnusson, L., Gabrielsson, E., von Platen, F., Hammarlund, J. and Wennerstein, B. (2000) *Från byggsekt till byggsektor* [From Construction Sect to Construction Sector]. Byggkostnadsdelegationens slutbetänkande [Final reflections from the Construction Cost Delegation], SOU 2000:44, Fritzes, Stockholm.

Yoon, J.H. and Kang, B.G. (2003) The drivers for change in the Korean construction industry: regulation and deregulation. In G. Bosch and P. Philips (eds) *Building chaos: an international comparison of deregulation in the construction industry*, Routledge, London, 210–33.

Yu, A.T.W., Shen, G.Q.P., Kelly, J. and Hunter, K. (2006) *A How-To Guide to Value Briefing.* Research monograph, The Hong Kong Polytechnic University, Hong Kong.

Zaheer, A. and Venkatraman, N. (1995) Relational governance as an interorganizational strategy: an empirical test of the role of trust in economic exchange. *Strategic Management Journal*, 16(5), 373–92.

Zayed, T.M. and Halpin, D.W. (2004) Simulation as a tool for pile productivity assessment. *J. Construction Engineering and Management*, 130(3), 394–404.

Zhang, H. and Tam, C.M. (2005) Consideration of break in modeling of construction processes. *Engineering, Construction and Architectural Management*, 12(4), 373–90.

Zollo, M. and Winter, S.G. (2002) Deliberate learning and the evolution of dynamic capabilities. *Organization Science*, 13(3), 339–51.

Zucker, L.G. (1986) Production of trust: institutional sources of economic structure. In B.M. Staw and L.L. Cummings (eds) *Research in Organizational Behaviour*, 8, 53–111.

Index

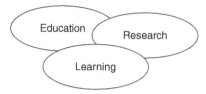

Figure 1.3 Means for supporting change.

programs, courses and tuition. Education can be delivered in dedicated organizations such as schools, colleges and universities, but also within business organizations – corporate education can be regarded as a means for filling gaps in tertiary education. Learning is a concept that focuses on the needs of an individual. It is possible to develop an educational program, but the participant (i.e. the actor) is the only one who can learn. Research is a concept that focuses on a structured process for developing new knowledge that can be used in education and as part of individual learning.

Operational level

In order to develop the construction field, one must recognize that the desired changes are closely linked to individuals. If changes happen in the construction process, the individual is likely to be affected in some way. It can be values in relation to the construction process and collaboration between actors that have to be adjusted. Appropriate behavior on site and in the factory is critical for the success of the construction process and it may be necessary to develop this aspect much further so that new patterns and levels of work performance can emerge. Attitudes toward end-users, quality and safety on the site must be openly discussed and changed where required. Changes at the operational level have also to do with the techniques and tools that are required in the construction process.

Research and researchers in the field

Three key concepts underscore the contribution that a researcher can make to both a *good* research process and *good* research results. The concepts are shown in Figure 1.4.

In order to produce research that can deliver new knowledge, it is necessary to have well-educated researchers. First of all, researchers need deep knowledge and understanding of the relevant theories as they apply to their research area. The researcher also needs a sound grasp of methods that can be used to confirm and then pursue relevant research questions. Most research projects in the field of construction require reliable and validated empirical data. A conclusion is that the researcher must continue to learn about these three concepts and use them throughout the whole